Perception, Theory and Commitment
The New Philosophy of Science

Harold I. Brown

Perception, Theory and Commitment
The New Philosophy of Science

Precedent Publishing Inc.

Chicago 1977

Q
175
,B796

Contents

Acknowledgements

I wish to express my gratitude to a number of individuals and groups who helped bring this book into existence and improve it. My good friend Lester Embree's suggestion that I write a (rather different) book on the new philosophy of science started me thinking along the lines that led to this book. The central thesis of the book and many of the arguments were first presented to the members of my course on Contemporary Philosophy of Science during the Fall, 1971 semester at Northern Illinois University and they helped improve a number of the arguments as well as eliminate some poor ones. I have discussed many of these issues with Theodore Kisiel, and David Stein read and commented on the entire manuscript. Henry Cohen provided much valuable service as consulting editor and critic. I am solely responsible for the errors that remain.

I am also deeply indebted to the Dean's Fund at Northern Illinois University for a research stipend during the summer of 1972 and to the National Endowment for the Humanities for a Younger Humanist Fellowship, 1972-73. The secretarial staff of the Philosophy Department at Northern Illinois University typed and retyped the manuscript with the best of humor and the Philosophy Department and Dean's Fund provided funds for duplicating the manuscript.

Introduction

Throughout the first half of the twentieth century philosophy of science was dominated by logical empiricists who took classical empricism and the powerful tools of modern symbolic logic as the basis for their analyses of science. Philosophers working in this tradition concerned themselves primarily with logical problems, particularly the logical structure of theories, and the logical relations between statements which describe observations and the laws and theories that these statements confirm or refute. Questions which are not amenable to formal analysis, such as the nature of scientific discovery, were brushed aside as nonphilosophical. Similarly, logical empiricists did not concern themselves with the nature of scientific progress, although they tended to accept the traditional view that modern science came into existence in the sixteenth and seventeenth centuries with the discovery of the "empirical method" and has exhibited a history of steady accumulation of knowledge.

Since the 1950's the methods and conclusions of logical empiricism have been under sustained attack by a number of writers of rather diverse philosophical backgrounds. Among the founding works of the new approach, Norwood Russell Hanson's *Patterns of Discovery* and Michael Polanyi's *Personal Knowledge* appeared in 1958, Stephen Toulmin's *Foresight and Understanding* in 1961, and Thomas S. Kuhn's *The Structure of Scientific Revolutions* and Paul K. Feyerabend's essay "Explanation, Reduction, and Empiricism" in 1962. In contrast to logical empiricism, the outstanding feature of the new approach is a rejection of formal logic as the primary tool for the analysis of science, and its replacement by a reliance on detailed study of the history of science. Although there are many disagreements among advocates of the new approach, there are enough common themes to justify talking about a "new image of science."

Most scientific research consists, in this view, of a continuing attempt to interpret nature in terms of a presupposed theoretical framework. This framework plays a fundamental role in determining what problems must be solved and what are to count as solutions to these problems; the most important events in the history of science are revolutions which change the framework. Rather than observations providing the independent data against which we test our theories, fundamental theories play a crucial role in determining what is observed, and the significance of observational data is changed when a scientific revolution takes place. Perhaps the most important theme of the new philosophy of science is its emphasis on continuing research, rather than accepted results, as the core of science. As a result, analysis of the logical structure of completed theories is of much less interest than attempting to understand the rational basis of scientific discovery and theory change.

The present book has two aims. The first is to survey the main themes of the new philosophy of science, to develop them further, and to attempt to resolve some of the problems raised by this approach. The second is to argue that the picture of research controlled by a body of presuppositions applies to the philosophy of science as well as to science. I will attempt to show in detail that logical empiricism was a research project of this sort, and that the development of the new philosophy of science constitutes an intellectual revolution in philosophy of the same kind as scientific revolutions.

Part I of the book deals with logical empiricism. The first chapter is an outline of the philosophical background of logical empiricism. This consists essentially of the Humean version of classical empiricism, modified and developed as a result of modern symbolic logic and logical positivism. The next three chapters are studies of central problem areas of logical empiricist research: confirmation, theoretical terms, and explanation. The emphasis here is on showing in detail how acceptance of Humean empiricism and symbolic logic as the basis for a philosophy of science has controlled this research, and how failures to solve the problems generated have led to continual modification of the logical empiricist project. The final chapter of Part I is an analysis of Popper's falsificationism as a transition view between logical empiricism and the new approach.

Part II takes up the central themes of the new philosophy of science: the relation between perception and theory, the role of presuppositions in scientific research, the nature of scientific revolutions, and the nature of discovery and scientific progress. In the last chapter I examine the epistemology which is implicit in the new image of science. This reversal of the order of Part I is necessary because logical empiricism is a philosophy of science which resulted from an attempt to apply an already well developed epistemology to modern science, while the new movement emerged largely as a reaction to the failure of logical empiricism to carry through its own research program. As a result, the alternative epistemology of the new movement must be dug out of its analyses of problems in the philosophy of science.

Each of the two parts can stand, to a large degree, by itself and the reader who is particularly interested in an overview of the new philosophy of science can go directly to Part II. At the same time they supplement each other in important ways. Part I provides the historical background for Part II; Part II provides the analysis of science which Part I extends to the philosophy of science. And the two parts together constitute one argument for the thesis that there are fundamental similarities between scientific method and philosophic method.

Part I

Logical Empiricist
Philosophy of Science

The Origins of
Logical Empiricism

Our concern in Part I is to review some of the central problems
of logical empiricist philosophy of science and to examine the
relation between these problems and the theory of knowledge that
is presupposed by logical empiricist analyses of science. We will
formulate the dominant themes of this theory of knowledge by
tracing the main stages in its development. Our starting point will
be the Humean version of classical empiricism and we will then
examine how this empiricism was modified by the development of
modern symbolic logic and the work of the logical positivists.
Since our concern here is not with Hume himself, but with the
philosophical framework of logical empiricism, we will summarize
the interpretation of Hume that has influenced the development
of this framework.

Humean
Empiricism

The two central problems of the theory of knowledge are the
problems of meaning and of truth, and the empiricist approach
to these problems received its classical form in the work of
David Hume. Hume's approach can be developed most clearly
in terms of a three-fold distinction between impressions, ideas
and language. Book One of the *Treatise of Human Nature* begins
with the statement, "All the perceptions of the mind resolve
themselves into two distinct kinds, which I shall call IMPRESSIONS
and IDEAS."[1] Impressions are the immediate objects of aware-

ness that we experience when we perceive or introspect. Ideas are the objects we are aware of in all mental activities other than perception and introspection, e.g., whenever we reflect, remember, imagine, and so forth, and we must distinguish two kinds of ideas, simple ideas and complex ideas. Simple ideas are copies of impressions which remain in the mind after an impression has occurred and which differ from impressions only in that they are less forceful and lively. Complex ideas are ideas that the imagination creates by combining simple ideas. The imagination can bring together any set of simple ideas to form a complex idea, but it can create no new simple ideas; thus the range of ideas I can entertain is limited by the range of impressions I have experienced.

Impressions and ideas provide, for Hume, a complete inventory of objects of awareness, but they do not constitute knowledge. All knowledge is formulated in propositions and it is with respect to propositions that the two central questions of epistemology—how we determine if a purported proposition is meaningful and how we determine which meaningful propositions are true—arise The basic unit of meaning, for Hume, is the term, and a term is meaningful only if there is an idea which corresponds to it. An individual can learn the meaning of a term only if he has experienced the impressions necessary for the formation of the corresponding idea, and any term which is supposed to refer to an object beyond the limits of possible experience is a mere meaningless sound or mark. A purported proposition which contains a single meaningless term is itself a meaningless pseudo-proposition and is neither true nor false. Thus the range of meaningful language is limited to the range of possible experience.

Meaningful propositions must be further subdivided into two kinds, relations of ideas and matters of fact. As the name suggests, statements of relations of ideas assert connections which hold between ideas, their truth-value being determined solely by reflecting on these ideas. Knowledge of relations of ideas is a priori, and is the only form of a priori knowledge which Hume will admit; all true statements of relations of ideas are necessary truths and all false statements of relations of ideas are self-contradictory. Statements of matters of fact refer to the experienced world and the truth-value of such statements is determined by reference to experience. Every statement of a matter of fact is ultimately equivalent to a set of assertions about what kinds of impressions occur in conjunction with each other, and we test these statements by observing the occurrence or non-occurrence of these impressions.

Besides being the source of meaning and truth, impressions, for Hume, are also the ultimate existents, the fundamental building blocks of reality. The only world that can be known is the world of impressions and every impression is ontologically distinct from every other impression, i.e., the existence or nonexistence of any impression is completely independent of the existence or nonexistence of any other impression. This thesis, however, raises important problems about the nature of our knowledge of the experienced world. Suppose that I have observed that a particular set of impressions always occur together, that, for example, a particular species of color, odor, shape, etc. which I call "fire" has always occurred in conjunction with an impression of heat (from an appropriate distance). According to Hume, there is no connection between the impression of heat and the other impressions in this set, thus I have no logically adequate reason for assuming that these impressions will occur together in the future. From the point of view of the philosophy of science, this raises the central issue of the grounds for accepting any universal proposition. Every universal proposition entails predictions about future experience but if there is no necessary connection between impressions that have occurred together in the past, then there is no guarantee that they will continue to occur together in the future. A similar problem is raised from the viewpoint of everyday life, since our day to day survival depends on the assumption that future experience will follow the same patterns as past experience. Hume resolves the latter problem by means of a psychological account of how we form the habit of expecting the future to resemble the past and how we act in accordance with this habit. This approach is not adequate for the purposes of the philosopher of science, whose problem is to find the rational grounding for the acceptance of universal scientific laws any attempt to replace such a rational grounding with mere habits would amount to a rejection of the rationality of science. Thus, as we shall see in some detail, the problem of how universal laws are empirically confirmed has remained one of the central research problems for empiricist philosophy of science.

Historically, one of the most important objections to empiricism has come from the philosophy of mathematics. In mathematics, and particularly in mathematics as applied in science, we seem to have a body of knowledge which is relevant to matters of fact but which is known a priori. It is not by experience that we have come to know that $2+3=5$ or that the enclosed angles of all Euclidean plane triangles add up to 180 degrees, yet scientists apply mathematics to experience with great success. We have already seen that the only a priori knowledge Hume will allow is knowledge of relations of ideas, but Hume's concept of a relation of ideas is not sufficiently clear and well developed to serve as the basis for a full blown philosophy of mathematics. The degree to which Hume himself was troubled by this problem can perhaps be indicated by noting that in the *Treatise,* where the distinction between "matters of fact" and "relations of ideas" is not explicitly drawn, he holds that only arithmetical and algebraic reasoning are capable of achieving certainty and that geometry, which deals with sensed qualities, is inexact.[2] In the *Enquiry Concerning Human Understanding,* Hume draws the distinction between matters of fact and relations of ideas and holds that arithmetic, algebra and geometry consist of relations of ideas and are all exact and certain.[3] It is not until the twentieth century, with the development of modern symbolic logic and the logicist theory of mathematics, that we approach an analysis of mathematics which is adequate from an empiricist standpoint.

Logicism

The central thesis of the logicist position is formulated by Russell in the preface to *The Principles of Mathematics:* "That all pure mathematics deals exclusively with concepts definable in terms of a very small number of fundamental logical concepts, and that all its propositions are deducible from a very small number of fundamental logical principles. . . ."[4] The complete proof of this thesis is attempted by Whitehead and Russell in the three volumes of *Principia Mathematica,* and in order to carry out their argument Whitehead and Russell developed a powerful new form of logic. We must examine the structure of this logic.

The central feature of principia logic is that it is an extensional logic; in particular, in the case of propositional logic it is truth-functional. A distinction is drawn between "elementary" or "atomic" propositions and "molecular" propositions, where molecular propositions are constructed out of elementary propositions by means of operators. Elementary propositions are either true or false and the propositional operators are so defined that the truth-value of a molecular proposition is uniquely determined by the truth-values of its constituent elementary propositions. At no point in the evaluation of molecular propositions does the meaning or content of the constituent propositions play any role. For example, the conjunction of two propositions, p and q, is true whenever both p and q are true, and false otherwise. Thus, within the structure of principia logic there is no significant difference between conjoining two propositions which refer to the same subject matter, such as "Electron e is in a gravitational field" and "Electron e is in a magnetic field," and conjoining two statements which have no subject matter in common, such as one of the above statements and "George Washington was born on February 22." This feature of principia logic has had, as we shall see, a significant influence on the work of the logical empiricists who have adopted principia logic as their primary tool for the analysis of science.

The attempt to construct a truth-functional interpretation for all propositional operators becomes particularly problematic in the important case of implication. The demand for a completely truth-functional propositional logic requires that "$p \quad q$" have a truth-value for every combination of truth-values of p and q—including the case in which the antecedent p is false. Within the context of the philosophy of mathematics this problem can be handled quite neatly,

The essential property that we require of implication is that, "What is implied by a true proposition is true." It is in virtue of this property that implication yields proofs. But this property by no means determines whether anything, and if so what, is implied by a false proposition. What it does determine is that, if p implies q, then it cannot be the case that p is true and q is false, i.e., it must be the case that either p is false or q is true. The most convenient interpretation of implication is to say, conversely, that if either p is false or q is true, then "p implies q" is to be true. Hence "p implies q" is to be defined to mean, "Either p is false or q is true."[5]

The definition of "$p \supset q$" as logically equivalent to "$\sim p \vee q$" has the effect of assigning the value of "true" to any implication with a false antecedent. This may seem strange, but it creates no difficulties at all for the philosopher of mathematics since in mathematics we are only concerned with formal proofs which take place from premises assumed to be true. As Whitehead and Russell point out in the passage quoted above, the essential property of implication for the mathematician and the philosopher of mathematics is that anything implied by a true proposition must be true. In his *Introduction to Mathematical Philosophy* Russell again emphasizes the same point: "In order that it may be *valid* to infer q from p, it is only necessary that p should be true and that the proposition 'not-p or q' should be true. Whenever this is the case, it is clear that q must be true."[6]

While this notion of "material implication" may be quite adequate for the analysis of mathematical inference, logical empiricists have extended the use of the principia formalism well beyond the limits of pure mathematics. This is particularly true for the principia analysis of universal affirmative propositions, propositions of the form "All P are Q." The symbolic analysis of this proposition form is based on noting that "All P are Q" is logically equivalent to the hypothetical "If anything is P, then it is Q." "All P are Q" is symbolized as "$(x)(Px \supset Qx)$" and this analysis is built on the notion of material implication, so that even though "$(x)(Px \supset Qx)$" is not itself truth-functional, the properties of material implication have been encapsulated in it. In this way the properties of material implication also become encapsulated into the analysis of scientific laws such as "All ravens are black," "All electrons have negative charge," or "All chemical reactions between an acid and a base yield water and a salt" when "$(x)(Px \supset Qx)$" is taken to be an adequate formulation of these laws.

Let us now consider the logicist solution to the problem of mathematical truth. Logicism maintains that mathematics is logic and thus that mathematics is true in just those ways in which logic is true. Unfortunately, as Russell himself pointed out, this does not solve the problem of the nature of mathematical truth, but only transforms it into the problem of the nature of logical truth; for this problem Russell was able to offer no satisfactory solution. Russell sums up the problem in his *Introduction to Mathematical Philosophy*:

> It is clear that the definition of "logic" or "mathematics" must be sought by trying to give a new definition of the old notion of "analytic" propositions. Although we can no longer be satisfied to define logical propositions as those that follow from the law of contradiction, we can and must still admit that they are a wholly different class of propositions from those that we come to know empirically. They all have the characteristic which, a moment ago, we agreed to call "tautology." This, combined with the fact that they can be expressed wholly in terms of variables and logical constants (a logical constant being something which remains constant in a proposition even when *all* its constituents are changed)—will give the definition of logic or pure mathematics. For the moment, I do not know how to define "tautology."[7]

Appended to this passage is the following footnote, "The importance of 'tautology' for a definition of mathematics was pointed out to me by my former pupil Ludwig Wittgenstein, who was working on the problem. I do not know whether he has solved it, or even whether he is alive or dead."[8]

Russell's *Introduction to Mathematical Philosophy* was published in 1919; two years later Wittgenstein's *Logisch-Philosophische Abhandlung* appeared and the following year an English translation came out under the title *Tractatus Logico-Philosophicus*. In the *Tractatus* Wittgenstein introduced truth-tables and used them as a basis for formulating a definition of "tautology" which has become standard among logical empiricists. Truth-tables provide a mechanical method for computing all possible truth-values of a molecular proposition. When the complete truth-tables for various proposition forms are constructed we find that they divide into three types: forms which are true for some values of the arguments and false for others, forms which are false for all values of the arguments, and forms which are true for all values of the arguments. It is this last class which Wittgenstein calls "tautologies" and which includes all logical and, for the logicist, all mathematical truths. The thesis that all logical truths are tautologies is clearly consistent with traditional definitions of logical truth such as "True in all possible worlds," or "True by virtue of the form alone," as well as with the demands of empiricism. Tautologies say nothing about the world but only about our use of symbols, so that the empiricist need have no qualms about admitting them as true a priori. When used in conjunction with empirical propositions in logical or mathematical reasoning, tautologies provide a means of transforming empirical propositions into other empirical propositions without changing their truth value; it is from this property of tautologies that their usefulness for science derives.

There is another approach to the philosophy of mathematics that should be mentioned here because it is a close cousin of logicism and has had an equal impact on modern empiricists:

Hilbert's *formalism*. For the formalist, pure mathematics, including logic, consists of uninterpreted calculi, axiom systems which are to be manipulated by means of a set of formal rules or algorithms. As in the case of logicism, for the formalist pure mathematics says nothing about the world, but while the logicist holds that pure mathematics and logic are true, the formalist holds that they are neither true nor false, being merely rule-governed games with symbols. Mathematics may be applied to scientific problems by giving appropriate interpretations to the symbols, but once this is done we are dealing with applied mathematics and the question of the acceptability of any system of applied mathematics for a particular area of scientific research becomes an empirical question. For both the formalist and the logicist logic is concerned solely with syntax, i.e., with formal relations between symbols, and all arguments must consist of the manipulation of symbols in accordance with precise rules. This identification of logic with syntax has been a major feature of logical empiricist studies of the logic of science. Empiricism and the new symbolic logic were welded together and developed into a philosophy of science by logical positivism, to which we now turn.

Logical
Positivism:
the Vienna
Circle

The term "positivism," coined by Auguste Comte, in general is used as a name for a form of strict empiricism: the positivist maintains that only those knowledge claims which are founded directly on experience are genuine. Modern logical positivism, in particular the positivism of the Vienna Circle, is a form of positivism which accepts the symbolic logic of *Principia Mathematica* as its primary tool of analysis. For the logical positivist there are two forms of research which yield knowledge: empirical research, which is the task of the various sciences, and logical analysis of science, which is the task of philosophy. We will take Wittgenstein's *Tractatus* as the central source for our discussion of logical positivism since it was hailed as such by the members of the Vienna Circle. It should be noted, however, that the correct interpretation of many of Wittgenstein's pronouncements in the *Tractatus* is rather controversial and it is not my intention to enter that controversy here, but only to present the interpretation of Wittgenstein that was accepted by the Vienna Circle.

The central doctrine of logical positivism is the *verification theory of meaning,* the thesis that a contingent proposition is meaningful if and only if it can be empirically verified, i.e., if and only if there is an empirical method for deciding if it is true or false; if no such method exists it is a meaningless pseudo-proposition. In order to understand the full thrust of this thesis it will be helpful to set it in the context of Wittgenstein's notion of "facts."

For Hume the basic elements of experience are impressions; for Wittgenstein the basic units of experience are facts: not just qualities such as "red " but rather "that there is red at a given time and place." The significance of this distinction can be shown most clearly by restating it in principia notation. In this symbolism Hume's impressions would be symbolized by a predicate such as "P," a fact, on the other hand, is an individuated predicate and thus would be symbolized "Pa." For Wittgenstein, as for Hume, the fundamental unit of meaningful language must correspond to the fundamental unit of experience; while for Hume the fundamental unit of meaning is the term which refers to an idea, for Wittgenstein it is the atomic proposition which refers to an atomic fact.

A number of Hume's central doctrines are now carried over into the *Tractatus*. For Hume impressions are the ultimate existents and for Wittgenstein atomic facts play this role. Thus Wittgenstein writes: "The world is the totality of facts, not of things."[9] and "The world divides into facts."[10] And while for Hume every impression is distinct from every other impression and the only necessity is the logical necessity of relations of ideas, so for Wittgenstein "Each item can be the case or not the case while everything else remains the same,"[11] and "There is no compulsion making one thing happen because another has happened. The only necessity is *logical* necessity."[12] Similarly, for Wittgenstein the atomic propositions which make up the fundamental layer of our empirical knowledge are all logically distinct (as for Hume simple ideas are all logically distinct). No atomic proposition can be deduced from another atomic proposition, nor can any one atomic proposition contradict another. "The simplest kind of proposition, an elementary proposition, asserts the existence of a state of affairs,"[13] and "It is a sign of a proposition's being elementary that there can be no elementary proposition contradicting it."[14] Our empirical knowledge, then, ultimately consists of a set of elementary propositions within which any propositions can be changed without this having any effect on any other propositions.

Fundamental to Wittgenstein's argument is a further distinction between "facts" (*tatsache*), "The world divides into facts"[15] and "states of affairs" (*Sachverhalt*), "A state of affairs (a state of things) is a combination of objects (things)."[16] A state of affairs is a logically possible fact, a fact is a state of affairs which happens actually to be the case. Any proposition which corresponds to a state of affairs is meaningful, a proposition which corresponds to a fact is also true, and a proposition and the state of affairs to which it refers have the same logical form. A meaningful proposition is a logical picture of a state of affairs and in a logically correct language all meaningless combinations of words, all pseudo-propositions, will violate the syntactical rules of the language. Needless to say, no naturally existing language meets these conditions. One of the central concerns of the logical positivist is the construction of such a logically correct language, and it should come as no surprise that the logical formalism of *Principia Mathematica* is taken as the basis for the construction of this language.

We can now return to the verification theory of meaning and clarify what is meant by the strict positivist notion of verification. In order to do this we will divide purported propositions into four kinds: First, there are purely formal propositions, tautologies and contradictions. These are meaningful and we determine their truth-value by examining their form. Second, there are atomic propositions. These are also meaningful and we determine their truth-value by observing whether they conform or fail to conform to the facts. Third, there are molecular propositions. These are truth-functions of atomic propositions and their truth-value is determined by first determining the truth-values of the constituent atomic propositions and then applying the definitions of the logical constants. Lastly, there are all other combinations of words which do not fall into any of the above classes. These are pseudo-propositions, mere meaningless combinations of sounds or signs without cognitive content. Thus the truth-value of any meaningful proposition can be determined once and for all solely by means of observation and logic.

Logical Empiricism

Logical empiricism can best be understood as a more moderate version of logical positivism. The central difficulty for logical positivism as a philosophy of science is that scientific laws which are formulated as universal propositions cannot be conclusively verified by any finite set of observation statements. Some of the members of the Vienna Circle, such as Schlick and Waismann, accepted this conclusion but avoided having to relegate scientific generalizations to the realm of meaningless pseudo-statements by maintaining that they are not propositions at all, but rules which allow us to draw inferences from observation statements to other observation statements. But most of the positivists chose instead to give up the strict verificationist theory of meaning and replace it with the requirement that a meaningful proposition must be testable by reference to observation and experiment. The results of these tests need not be conclusive, but they must provide the sole ground for determining the truth or falsity of scientific propositions. We can identify the beginnings of logical empiricism with this liberalization of the logical positivist theory of meaning; indeed, we can be somewhat more specific than this, for Rudolph Carnap's *Testability and Meaning*[17] can reasonably be viewed as the founding document of logical empiricism.

Carnap acknowledges the impossibility of conclusively verifying any scientific proposition. He proposes replacing the notion of verification with the notion of a "gradually increasing confirmation,"[18] and takes the notion of an "observable predicate" as fundamental, defining "confirmable sentence" in terms of this notion.[19] The effect of this latter move is a rejection of the positivist thesis that the sentence is the fundamental unit of meaning and a return to the older Humean concern with the meaning of terms. Thus two of the central problems of logical empiricist philosophy of science are the analysis of the confirmation relation that is to hold between a scientific law and the observation statements which confirm or disconfirm it, and the analysis of how scientific terms get their meaning. The latter problem is particularly pressing for the empiricist in the case of the theoretical terms of modern physics, terms such as "electron," "entropy," and "state-function," since these terms do not appear to refer to observables. We will begin our examination of logical empiricist philosophy of science with the problem of confirmation.

Chapter Two

Confirmation

The problem of confirmation can be looked at as a quantitative problem or as a qualitative problem, although these two viewpoints are by no means mutually exclusive. Quantitative theories of confirmation attempt to assign a numerical degree of confirmation to hypotheses on the basis of observational evidence; qualitative analyses of confirmation are concerned with the problem of explicating the relation between an hypothesis and the observation reports which confirm it. The latter problem is logically prior in that we must be able to recognize which instances stand in a confirming or disconfirming relation to an hypothesis before we can attempt to quantify that relation.[1] Our discussion here will be confined to the qualitative problem.

The Paradoxes
of Confirmation

The classic study of the problem of defining "confirmation," which has served as the basis for most of the subsequent discussion, is Carl Hempel's "Studies in the Logic of Confirmation."[2] Hempel formulates his purpose as follows: "It ought to be possible, one feels, to set up purely formal criteria of confirmation in a manner similar to that in which deductive logic provides purely formal criteria for the validity of deductive inference."[3] We will not be concerned in this chapter to defend or attack the various attempts by philosophers to offer an acceptable definition of confirmation, but rather to analyze the way in which the logical empiricist formulation of the problem is controlled by the structure of the philosophical framework within which they are working. We have already seen that *logical* empiricism and *logical* positivism received much of their impetus from the new logic of Whitehead and Russell's *Principia Mathematica*; it should be clear that in setting as his desideratum the construction of a purely formal analysis of confirmation, Hempel is taking the achievement of *Principia Mathematica* as his model and attempting to extend its techniques into a new area.

From this point of view, logical empiricism is most fruitfully viewed as being not a body of doctrine, but rather a *research program.*[4] The philosophers who are engaged in this program begin with a common set of intellectual tools and techniques, and use these tools and techniques as the means of analyzing the nature of scientific knowledge. But it is the attempt to analyze science in terms of this particular philosophical framework that generates the particular body of problems with which logical empiricist philosophers of science have been concerned and determines the kinds of solutions that have been considered acceptable, the kinds of objections that are raised against proposed solutions from *within* the logical empiricist camp, and the general development of the logical empiricist problematic. Thus Hempel not only takes the formal structure of principia logic as his model in attempting to analyze confirmation, he also formulates the problem in the symbolism of principia logic. As we shall see, at least some of the problems he must resolve derive from the structure of this logic.

Hempel begins his discussion with a highly plausible proposal due to Nicod: for a given scientific law of the form "$(x) (Px \supset Qx)$" any observation sentence of the form "$Pa \cdot Qa$" is a confirming instance while any observation sentence of the form "$Pa \cdot \sim Qa$" is a disconfirming instance.[5] But in spite of its initial plausibility, the proposal leads directly to a difficulty. For the proposition "$(x) (\sim Qx \supset \sim Px)$" is logically equivalent to "$(x) (Px \supset Qx)$" but will have, on Nicod's criterion, different confirming instances since only "$\sim Qa \cdot \sim Pa$" will confirm "$(x) (\sim Qx \supset \sim Px)$" according to this criterion. "This means that Nicod's criterion makes confirmation depend not only on the content of the hypothesis, but also on its formulation,"[6] i.e., logically equivalent sentences presumably say the same thing, and the question of whether a particular observation statement confirms a given hypothesis should depend only on the content of the hypothesis, not on its formulation. Indeed, Hempel points out, further symbolic manipulation shows that "$(x) (Px \supset Qx)$" is also logically equivalent to "$(x) [(Px \cdot \sim Qx) \supset (Rx \cdot \sim Rx)]$." But this sentence can have no confirming instances since nothing can satisfy both its antecedent and its consequent.[7]

Hempel attempts to resolve this difficulty by proposing as a general criterion which any adequate definition of confirmation must meet the *equivalence condition:* "Whatever confirms (disconfirms) one of two equivalent sentences, also confirms (disconfirms) the other."[8] But while this might seem to resolve the problem, it has the effect of generating a new difficulty which Hempel calls "the paradoxes of confirmation."[9] For if we accept the equivalence condition, we must accept "~ Qa · ~ Pa" as a confirming instance of "(x) (Px ⊃ Qx)." If, for example, the hypothesis in question is "All ravens are black," the discovery of any instance of an object which is not black and not a raven, such as a yellow pencil, would serve to confirm our hypothesis. And we can go a bit further: "(x) (Px ⊃ Qx)" is also logically equivalent to "(x)[(Px ∨ ~ Px) ⊃ (~ Px   Qx)]" and since the antecedent of this proposition is satisfied by everything (being a tautology), it follows that anything that is ~ P or Q also confirms "(x) (Px ⊃ Qx)." In terms of our raven example, the discovery of any object that is either not a raven, such as a pencil, or is black, such as a lump of coal, confirms the hypothesis that all ravens are black. As Goodman points out,[10] this opens up wonderful prospects for indoor ornithology since I can now, without ever leaving my study, accumulate innumerable confirming instances for the hypothesis in question or indeed for any hypothesis that can be formulated in the form "(x) (Px ⊃ Qx)."

There have been many attempts to resolve these paradoxes. R. G. Swinburne,[11] in a recent survey of the literature, has divided the attempted solutions into three types: those which reject the adequacy of "(x) (Px ⊃ Qx)" as a formulation of universal scientific laws; those which reject the equivalence condition; and those which reject Nicod's criterion. Within this last class there are, again, a number of different sub-types. One approach, for example, adopted by the Popperians Watkins and Agassai, consists of adding an additional criterion which an observation sentence must meet before it can count as a confirming instance, i.e., it must have occurred in the process of testing the proposition in question. Another approach adopted by Hosiasson-Lindenbaum, Pears and Alexander rejects the possibility of a qualitative analysis of confirmation and approaches the problem in terms of the relative size of the various classes involved and the effects of this parameter on the degree of confirmation. In the midst of this literature, however, Hempel's own proposal (which falls into this third class) remains in many ways the most interesting. Hempel rejects Nicod's proposal that only observation statements of the form "Pa · Qa" or "Pa · ~ Qa" are relevant to the confirmation or disconfirmation of universal propositions of the form "(x) (Px ⊃ Qx)," accepts the consequences of the equivalence condition, and argues that, logically speaking, there is nothing paradoxical about the "paradoxes of confirmation" at all, but rather the appearance of paradox is merely a "psychological illusion."[12]

Hempel offers two arguments for this thesis. First[13] he argues that the appearance of paradoxicality derives from the mistaken assumption that the universal statement "All ravens are black" is only about ravens. Rather, Hempel maintains, it is about all of space-time; it says that at no place and at no time will we ever encounter an object which is both non-black and a raven. Understood in this way, any discovery of an object which fails to be a non-black raven constitutes a confirmation of the hypothesis.

Hempel's second argument is that the appearance of paradoxicality also derives from a failure to observe a "methodological fiction"[14] which must be observed in the logical analysis of all cases of confirmation. In so far as we are attempting to analyze the *logic* of confirmation, we are only concerned to analyze the relation between an hypothesis and a specified body of evidence, so that in any particular case we must adopt the fiction that the piece of evidence in question is all the information we have. Thus Hempel argues that if I hold a piece of ice in a flame and note that it does not turn the flame yellow, it would seem paradoxical to take this as evidence for the hypothesis, "All sodium salts burn yellow." But this is only an apparent paradox deriving from my allowing the additional information that the object in the flame is ice to intrude illegitimately into the analysis. If I were to hold an unknown object in the flame, note that it did not turn the flame yellow and, on subsequent analysis, determine that it was not a sodium salt, then, Hempel maintains, there would be nothing paradoxical about taking this as evidence that "All sodium salts burn yellow." Carrying this analysis one step further, if I examine an object and discover that it is black, the only information I may take into account for purposes of a formal analysis of confirmation is that I have a black object before me. And since this bit of information confirms, as Hempel points out,[15] the hypothesis that all objects are black, it certainly confirms the much weaker consequence of this hypothesis "All ravens are black." Hempel goes on to state that "Other paradoxical cases of confirmation may be dealt with analogously"[16] and although Hempel does not do so, the point is worth clarifying. There is only one other paradoxical case which has not been covered by the above discussion: the case in which I find an object which is not a raven. Presumably what Hempel has in mind is that when I encounter an object which is not a raven, I confirm the proposition that there are no ravens and this, in conjunction with the currently accepted interpretation of universal propositions according to which a universal proposition is true if its subject class is empty, constitutes a confirmation of the hypothesis that all ravens are black. (Unfortunately, given this interpretation of universal propositions, the discovery of a non-raven also confirms the proposition "No ravens are black," but we will let this point pass.)

Now why should we accept the proposed methodological fiction? Practicing scientists do not accept it, and it would seem to be the height of foolishness actually to assume it in the process of testing a scientific hypothesis since it requires the scientist to ignore most of what he knows. If Hempel's methodological fiction is to be taken seriously, then there is no reason why a scientist should not pick up a piece of ice, hold it in a flame, pretend that he does not know that it is ice, observe that the flame is not yellow, then note that the object he is holding contains no sodium salts and announce a new confirmation of the thesis that all sodium salts burn yellow. Similarly, it would make perfect sense for me to observe a pencil on my desk, note that is is moving at a velocity slower than that of light, and conclude that I had confirmed Einstein's thesis that nothing moves faster than light (with the possible exception of tachyons, but I could also observe that the pencil is not a tachyon and thus confirm both that there are no tachyons and that all tachyons move faster than light). Clearly scientific research is not conducted in this manner. In planning an experiment to test an hypothesis, or in deciding whether a particular body of information is relevant to the truth of some hypothesis, the scientist does not pretend that the experimental result in question is the only piece of evidence he has; rather he brings to bear every bit of evidence that he can muster in order to draw his conclusions. It might be replied here that the methodological fiction is not being proposed for the scientist, but rather for the logician who is attempting to analyze the logical structure of scientific reasoning. But this proposal raises the further question of whether a logic of confirmation which takes Hempel's methodological fiction as one of its fundamental assumptions can throw any light on the structure of scientific thought. It is not at all obvious a priori that this is the case. Logical calculi, like mathematical calculi, can be constructed independently of the examination of any body of knowledge or experience, but the fact that a calculus exists is not a sufficient reason for assuming that it is an appropriate tool for the analysis of any particular area of human experience. Thus it may well be possible to construct a calculus of confirmation in which the formal relations between observation statements and various kinds of general sentences are analyzed with great care, but it may still turn out that this particular calculus is not at all appropriate for the formulation and analysis of the confirmation relation as it is found in scientific practice.

Indeed, one of the striking aspects of logical empiricist philosophy of science that we shall continually encounter is a consistent lack of detailed analysis of actual scientific theories or of examples of scientific research. Rather what we find, and the literature on the paradoxes of confirmation is typical in this respect, is the analysis of proposition forms, the construction of artificial languages and calculi, and the occasional illustration of these calculi by reference to simple empirical generalizations such as "All ravens are black" or "All sodium salts burn yellow" with the assumption that this will somehow elucidate the structure of science. Yet these are not examples of significant scientific propositions and it is by no means obvious that, even if we did develop a precise analysis of the confirmation relation between such propositions and their instances, this would apply in any way to the reasons for accepting or rejecting Newton's law of gravitation, the principle of general relativity or the Schrödinger equation.

This should not be taken, however, as a criticism of logical empiricism, but rather as a description of an intellectual phenomenon which requires an explanation, one which I believe is forthcoming if we view the above in terms of the presuppositional framework discussed in the first chapter. Just as scientific research cannot be carried out merely by collecting data, but rather requires some set of assumptions about how nature behaves to tell the researcher what data to collect and then how to interpret it,[17] so the philosophy of science could not proceed just by collecting information about what scientists do. Again some set of assumptions, in this case assumptions about the nature of knowledge and how its structure is best analyzed, is required before the philosopher can begin his examination of scientific knowledge. In the case of the earlier versions of logical empiricism there is, to begin with, the empiricist assumption that all scientific knowledge consists of generalization from experience. If this is the case, then the most complex scientific theory is ultimately reducible to some set of generalizations from experience and there is no need to analyze complex generalizations when simple ones will do as well. Thus, within the framework of his presuppositions, the logical empiricist need not consider the fact that he does not analyze complex scientific examples as a defect, but rather as a methodological virtue. It simplifies the discussion without, he will claim, eliminating any essential features of scientific knowledge.

Second, we have seen that the logical empiricists have accepted the logic of *Principia Mathematica* as the primary tool for the analysis of science and that once this commitment has been made, it is the proposition forms of *Principia* and their manipulation that become the primary subject of discussion. Again, this is by no means illegitimate. The decision to talk about science in the language of *Principia Mathematica* is in many ways parallel to the scientist's decision to talk about nature in the language of, say, Euclidean geometry or differential equations. Once such a decision is made, however, the structure of the accepted conceptual tools plays a central role in determining what phenomena the scientist or the philosopher focuses his attention on, in generating his problems, and in determining what kinds of solutions to these problems are taken to be acceptable. This point can be illustrated in the case of the paradoxes of confirmation, for these paradoxes might not arise at all, and certainly would not arise in as many forms, if we were working in terms of Aristotelian logic. Let us first consider the form of the paradoxes generated by the equivalence of "All ravens are black" and "All non-black things are non-ravens." If we accept an Aristotelian logic with the traditional analysis of existential import according to which a universal proposition asserts the existence of members of its subject class, then the two propositions in question are not equivalent since "All ravens are black" asserts the existence of ravens while "All non-black things are non-ravens" asserts the existence of non-black things; hence the paradox does not arise.

On the other hand, the acceptance of the traditional notion of existential import entails that all scientific propositions which refer to objects that do not actually exist are false, and since this includes just about all fundamental scientific propositions, it is a serious objection to the acceptance of this analysis of existential import. This has led at least one logician[18] to suggest that we reinterpret traditional logic as making no existence claims at all, and on this interpretation the paradox in question would arise even if we accepted an Aristotelian logic. However, the two remaining forms of the paradoxes that we have considered would not occur in any form of Aristotelian logic. One of these is that "All ravens are black" is confirmed by the discovery of any object which is either black or a non-raven because "(x) (Px ⊃ Qx)" is logically equivalent to "(x) [(Px v ~ Px) ⊃ (~ Px v Qx)]," and the other is that "(x) (Px ⊃ Qx)" is logically equivalent to "(x) [(Px · ~ Qx) (Px · ~ Px)]," which can have no confirming instances at all, but neither of these equivalences have any place in an Aristotelian logic.

The point here is not to argue the superiority of one form of logic over another, but only to point out that whether a particular problem arises at all depends on what form of logic we adopt for the logical analysis of science. Now just as alternative geometries are possible, so alternative philosophies of science are possible, and while the logical empiricists may or may not come to a mutually satisfactory resolution of the paradoxes of confirmation, for a philosophy of science which does not assume that all scientific statements must be formulable in the notation of principia logic, or which does not take the major task of philosophy of science to be the logical anlaysis of propositions, or which does not accept the assumption that scientific propositions receive their justification from some form of direct confirmation by experience, the paradoxes of confirmation are not significant problems. On the other hand, it must be emphasized that the continuing failure of the logical empiricists to reach agreement on a resolution of the paradoxes of confirmation—a problem which they themselves take to be a pressing one, to judge by the size of the literature it has generated—provides an important reason for looking seriously at alternative approaches to the philosophy of science.

Let us return now to Hempel's methodological fiction. Scientific practice not withstanding, this fiction is perfectly plausible once it is set in the context of the logical empiricist framework. For one of the assumptions of this framework is that the world is made up of distinct facts and that the propositions which describe these facts are all logically distinct from each other. Recall that for Hempel the paradoxes are a *psychological* illusion that results from taking into account information which ought, logically speaking, to be ignored. Within the framework in question, the fact that an object is, say, a piece of coal and the fact that it is black have nothing to do with each other, and the observation statements which describe these facts, "This is a piece of coal," and "This is black," are logically distinct statements.

We have seen how, according to Hempel, the observation of any black object can, logically speaking, confirm the statement "All ravens are black." But given this confirmation, the addition of another logically distinct proposition (that the object in question is a piece of coal and thus not a raven) can in no way alter the confirmation relation between the original observation statement and the hypothesis. It describes a distinct fact and is thus strictly irrelevant. To take into account the additional observation that the object I am examining is a piece of coal can, at most, cause psychological confusion, but it cannot affect the logical relation involved.

Another part of the logical empiricist framework conspires here to make the methodological fiction almost obvious. For, as we have seen, the paradigm on which Hempel is attempting to model his logic of confirmation is the deductive logic of *Principia Mathematica*. But it is a general principle of *deductive* logic that if a given set of premises entails a particular statement, then the addition of further statements to the premise set cannot affect this entailment relation. I submit that it is because he is attempting to analyze the confirmation relation with an eye on the deductive paradigm that Hempel assumes that if an hypothesis is confirmed by a given observation statement, the addition of further observation statements cannot negate that confirmation relation. In terms of the logical empiricists' research program, then, the methodological fiction provides a perfectly plausible approach to the resolution of the paradoxes of confirmation.

Confirmation and
Extensional Logic

We turn now to another problem in confirmation theory that has received some attention (although considerably less than the paradoxes) and which will allow us to illustrate further the role of the logical part of the logical empiricists' presuppositional framework in generating problems.

As a major goal of his study of confirmation Hempel develops a set of conditions of adequacy for any proposed definition of confirmation. One of these, the *special consequence condition,* states that "If an observation report confirms a hypothesis H, then it also confirms every consequence of H."[19] Now it might seem perfectly reasonable to include another condition which Hempel calls a *converse consequence condition*:[20] if an observation report confirms an hypothesis H, then it also confirms any other hypothesis which entails H; but the special consequence condition and the converse consequence condition taken together yield an unacceptable result. For if an observation statement O confirms H, then in accordance with the converse consequence condition it also confirms "$H \cdot G$" which entails H(G being any proposition at all). Application of the special consequence condition now yields the result that O confirms G. And since any observation report confirms some hypothesis, it follows that any observation report confirms any hypothesis.

Different writers have proposed different ways of resolving this problem. Barrett, for example, uses the above result as an argument against the acceptance of the special consequence condition[21] while Hempel, on the other hand, rejects the converse consequence condition[22] and Carnap rejects both the consequence condition (not just the special consequence condition) and the converse consequence condition.[23] But there is another possible source of this problem which none of these writers have considered, a source which lies at the very center of the logical empiricist program: the assumption that principia logic is an adequate tool for the analysis of scientific inference. We have already seen that one of the characteristic features of principia logic is that it is an extensional logic, that it does not take into account the *meaning* of the propositions involved in an argument. It is because of this extensionality that propositions such as "The charge on an electron is 1.6×10^{-19} coulombs and the earth is flat" are taken to be completely legitimate conjunctions from which one can infer either "The charge on an electron is 1.6×10^{-19} coulombs" or "The earth is flat."

Now the logical empiricist does not question the adequacy of principia logic as a tool for the analysis of science since the assumption of its adequacy is one of the defining characteristics of his research program, but it is only by assuming that any inference licensed by principia logic can be taken as a significant form of scientific reasoning that the above problem can be viewed as a problem for the philosophy of *science*. It is not at all difficult, however, to imagine a philosopher working from a different logical tradition, e.g., one which only permits the conjunction of propositions which have overlapping or common content. Such a philosopher might well be able to accept both the consequence and the converse consequence condition without being forced to draw the conclusion that any observation statement confirms every hypothesis. For such a philosopher, of course, the attempt to construct a purely syntactical definition of confirmation would never have become a research problem in the first place.

Goodman's Attack on Syntactical Analyses of Confirmation

The attempt to construct a purely syntactical definition of confirmation has been attacked by Nelson Goodman in a way that is particularly interesting for our purposes here. Taking Goodman's example, let us consider the proposition "All emeralds are green," and suppose that we have observed a large number of green emeralds and no instances of non-green emeralds. In terms of the kind of syntactical analysis that we have been examining, we have a proposition of the form "$(x) (Px \supset Qx)$" and a large number of observation statements of the form "$Pa \cdot Qa$;" this would appear to yield a highly confirmed generalization. But Goodman points out that, taking into account syntactical relations alone, it is by no means clear just what proposition has been confirmed. Consider, he suggests, the predicate "grue" defined as follows: "an object is grue if and only if it is green before time *t* and blue afterward." Since all of the observations made were made before *t*, as long as we restrict ourselves to syntactical criteria, we may well have a clear case of the confirmation of a universal proposition, but we have no grounds for determining if the proposition confirmed is "All emeralds are green" or "All emeralds are grue."[24] Goodman calls this problem "the new riddle of induction,"[25] and formulates it in terms of the concept of "projection."[26] Whenever we generalize or make any prediction on the basis of a given body of evidence, we are projecting that evidence into the future. The problem of induction (of which the problem of confirmation is the modern form) now becomes a case of the general problem of what sets of present evidence can be projected.

Goodman outlines his approach to the solution of this problem in the following passage:

> While confirmation is indeed a relation between evidence and hypotheses, this does not mean that our definition of this relation must refer to nothing other than such evidence and hypotheses. The fact is that whenever we set about determining the validity of a given projection from a given base, we have and use a good deal of other relevant knowledge. I am not speaking of additional evidence statements, but rather of the record of past predictions actually made and their outcome. Whether these predictions—regardless of their success or failure—were valid or not remains in question; but that some were made and how they turned out is legitimately available information.[27]

Let us attempt to clarify what Goodman is doing in this passage. To begin with we should note that he clearly accepts Hempel's methodological fiction.[28] But at the same time he is restricting its range of application by drawing a distinction between additional information which is *legitimately* available and which can thus be used, and additional information which cannot legitimately be used. Thus confirmation is still taken to be a relation between an hypothesis and an observation report and we may not invoke any further observation reports such as, to revert to one of our earlier examples, that the object I am holding in the flame is ice. But Goodman now tells us that there is another vast range of additional information that we can legitimately take into account: information about the history of past projections and their successes and failures. When I take this information into account, I find that as a matter of fact the predicate "green" has a long history of successful projections: there are numerous cases in which we have observed green objects and successfully generalized and made predictions from this observation. The predicate "grue" has no such history of successful projections. Goodman describes this situation by saying that the predicate "green" is much better "entrenched"[29] than "grue" and it is the degree of entrenchment of the predicate in question that provides us with a criterion for deciding between "All emeralds are green" and "All emeralds are grue." It is important to emphasize here that it is the historical record of actual projections and not the logical structure of the propositions involved that Goodman is appealing to. As Goodman himself is careful to point out, in any case in which "green" was projected, "grue" could have been projected;[30] it is the fact that we have been projecting "green" rather than "grue" that determines which of the two predicates is better entrenched.

Viewing Goodman's position in terms of the logical empiricist framework, we find that in taking the problem of induction to be a primary problem of the philosophy of science and in approaching this problem by an examination of simple generalizations, his work clearly exhibits some of the distinguishing characteristics of logical empiricism. On the other hand, his approach to a solution of the problem of induction also constitutes a significant break with the logical empiricist approach, since the main thrust of his argument is to reject the possibility of constructing a purely syntactical definition of confirmation. Goodman's work has generated a vast body of literature; it is not surprising to find that much of this has come from logical empiricists attempting to demonstrate that there is a clear syntactical difference between "green" and "grue" and thereby show that it is not only an historical fact that green has been projected in the past, but that there are sound logical reasons for doing so and for continuing to do so. One central point of attack, for example, has been the fact that the definition of "grue" contains a reference to a specific time while the definition of "green" does not.[31]

But there is another way in which we can interpret Goodman's new riddle of induction and his proposed solution, a way which suggests an even greater break with logical empiricism than the proposed abandonment of a syntactical analysis of confirmation. For what Goodman's attempt to solve the problem of projection in terms of the concept of entrenchment tells us, in effect, is that whether a given hypothesis has been confirmed by a particular set of observations can only be decided in the light of the history of science. Looked at in this way, Goodman's proposal raises questions which could be the subject of further research for philosophers of science who wanted to develop his position. How much of the history of science must we consider? The concept of a body moving to its natural place, for example, certainly has a longer history of actual projection than does the concept of gravitational attraction. Under what circumstances do new concepts such as the quantum of action appear and how do they become entrenched? How do previously well entrenched concepts lose their status? What is the relative importance of the collection of data in determining what concepts become entrenched as opposed to the importance of entrenched concepts in determining what data gets collected? And so forth.

What Goodman is proposing, then, is not a new solution to the problem of induction, but a new research program. We have seen at some length that the kind of approach to this problem which

Hempel's work exemplifies generates its own problems and that philosophers of science who wanted to develop his position. How tradition have been concerned to resolve these problems. Goodman, in offering a new way of looking at the problem of induction, generates a new set of problems; he implies that a new kind of research is necessary in order to solve these problems and that new criteria for the adequacy of a solution are in order. A philosopher who accepts Goodman's new approach and wishes to work on its problems cannot limit himself to logical analysis, but must engage in a good deal of historical research, and will surely find that criteria of formal adequacy are quite irrelevant to deciding whether a proposed solution to one of these new problems is acceptable. If enough philosophers take up the project of analyzing induction from the point of view of entrenchment, we will see the beginnings of a new research tradition in the philosophy of science. At the very least, a proposal which has the effect of suggesting to philosophers of science who are interested in induction that they change the questions they are asking and the kind of research that they are doing has emerged from within the logical empiricist tradition. This gives us some grounds for suspecting that the standard form of logical empiricist philosophy of science has lost a good deal of its vitality as a research program and that a new approach may be in order.

Chapter Three

Theoretical Terms

Explicit Definition   We have seen that one of the central doctrines of traditional
empiricism is that terms get their meaning as a result of having
been correlated with some set of impressions or, to use a more
contemporary term, "sense-data."[1] Any term which, in this view,
cannot ultimately be defined by reference to some set of
sense-data, has no meaning. Clearly, any philosopher who
examines modern physics from the point of view of this
presupposition (and who is not willing to reject physics as
meaningless) will find himself faced with a research problem. For
physics is full of terms which at least appear to refer to
non-observable entities, terms such as "electron,"
"state-function," and so forth, and the philosopher of science who
adopts an empiricist theory of meaning faces the problem of
showing how these terms can be defined by reference to
observables. A strict empiricist formulation of this research project
is given by Russell: "Wherever possible, logical constructions are
to be substituted for inferred entities."[2] Russell goes on to spell out
this project more completely:

> Given a set of propositions nominally dealing with the
> supposed inferred entities, we observe the properties which
> are required of the supposed entities in order to make these
> propositions true. By dint of a little logical ingenuity, we then
> construct some logical function of less hypothetical entities
> which has the requisite properties. This constructed function
> we substitute for the supposed inferred entities, and thereby
> obtain a new and less doubtful interpretation of the body of
> propositions in question.[3]

An example will help to clarify Russell's proposal. A large number of observed laboratory data can be accounted for by postulating that matter is made up of atoms which are themselves made up of smaller, more fundamental particles. The existence of discrete spectral lines, for example, can be explained by assuming that atoms include electrons which can exist only at discrete energy levels and that spectral lines are the result of radiation emitted when electrons drop from a higher to a lower energy level. Similarly, certain kinds of traces in cloud chambers, traces of a particular width and a particular curvature when the cloud chamber is in a magnetic field, can be explained as the result of the ionization of water droplets caused by the passage of an electron. And, to take one more example, the deflection of the needle of an ammeter which is part of a closed circuit can also be explained as a result of the passage of a stream of electrons through the circuit. In these cases (as well as the many other cases in which scientists make use of the notion of an electron), the electron is what Russell calls an "inferred entity." We do not actually perceive the electron, rather we infer its existence on the basis of observed data: lines on a spectrogram, streaks in a cloud chamber, the observed location of a pointer, etc. But since we do not perceive electrons, this leaves, for Russell, the problem of what we mean by the term "electron." According to Russell, we solve this problem by eliminating the inferred entity. Instead, we consider all the kinds of observation statements that are necessary in order to verify true propositions in which the term "electron" occurs, apply the techniques of symbolic logic to construct an appropriate logical function of these observation statements, and take this logical construction as a definition of "electron."

There have been two mainlines of argument against this approach from within the logical empiricist camp. One line, originally proposed by Ramsey[4] and developed in some detail by Braithwaite,[5] notes that if theoretical concepts are defined in the above manner, the theories in which these concepts occur will lose one of the most important functions of scientific theories. Contemporary empiricists generally agree that a central characteristic of successful scientific theories is that it is possible to predict new phenomena on the basis of them, and to use theories which were originally proposed in order to account for a particular range of phenomena to account for other phenomena that were not considered in their original construction. For example, Maxwell's electrodynamics predicted the existence of electromagnetic waves; and the molecular theory of gases, which was originally postulated to account for the pressure, volume, temperature relation of gases, also provided explanations for the law of diffusion of gases, for the specific heats of gases, and for many other phenomena. Indeed, a theory which was postulated in order to account for a specific set of phenomena and which could account for no other phenomena would not generally be considered a significant scientific achievement. But on Russell's approach, no theory would be extendable; for every time a new kind of data were taken into account, we would have to add this data into the definitions of our terms and we would thus be redefining the theoretical concepts rather than bringing a new area of experience under an old concept. This, it is argued, is not the way scientific theories are developed. Indeed, it is worthwhile to anticipate our argument for a moment and note that the stability of scientific concepts has become a central doctrine of many of the most recent logical empiricist writings. Some of the advocates of the new approach to the philosophy of science that we will examine in Part II maintain that one result of a scientific revolution is that scientific concepts change. Their logical empiricist opponents have maintained, in response, that the meanings of the terms remain unchanged while the truth-values of propositions in which they occur are changed.[6] Russell's thesis goes even farther than the so-called "radical meaning variance theorists," since for him every new empirical discovery, no matter how minor, forces us to redefine our concepts.

Before considering the second line of criticism referred to above, it will be convenient to examine briefly another attempt at explicit definition of theoretical concepts which has been highly influential: operationism. The operationist thesis, originally proposed by P. W. Bridgman, maintains that "In general, we mean by any concept nothing more than a set of operations; *the concept is synonymous with the corresponding set of operations.*"[7] Take, for example, the concept of length which Bridgman uses to introduce his approach. In order to specify what we mean by "length" we must specify the set of operations by which we determine the length of an object; this set of operations is the total meaning of the concept of length. This general thesis, for Bridgman, applies to all scientific concepts:

> If the concept is physical, as of length, the operations are actual physical operations, namely, those by which length is measured; or if the concept is mental, as of mathematical continuity, the operations are mental operations, namely those by which we determine whether a given aggregate of magnitudes is continuous.[8]

But consider a case in which we measure the distance between two points by two different methods: say by use of a measuring tape and by triangulation with a tape and a theodolite. The only operations required by the first method are operations of laying off a tape, but the second method requires not only laying off a tape, but also turning angles and computations. We are talking, then, about two different sets of operations and thus, for the operationist, not about two different ways of measuring length, but rather about two different concepts which, properly speaking, should be denoted by different terms.[9] There cannot, in principle, be two different ways of measuring a given parameter since, by hypothesis, the use of different methods of measurement entails that we are dealing with different parameters. Similarly, when physicists talk about nuclear dimensions of the order of 10^{-13} cm.

they are not talking about the same kind of thing that we talk about when referring to distances of one and two centimeters, nor are they talking about the same kind of thing that astronomers talk about when they discuss interstellar distances.[10] In each of these cases, the "distance" is determined by a different operation so that we are determining different things, and we should properly have three different words in our vocabulary.

Clearly operationism suffers from the same defect that we have already found in Russell's notion of explicit definition, but in a somewhat more extreme form: not only would operationism drastically limit the possibility of extending concepts into new areas, it would entail a great proliferation of the number of distinct theoretical concepts in contemporary science and the surrender of the goal of systematizing large bodies of experience by means of a few fundamental concepts.

Reduction Sentences

The second line of criticism of the demand for explicit definition of theoretical terms stems from Carnap's discussion of disposition terms.[11] Taking his example, suppose we attempt to define the term "soluble in water" in terms of observables in the following manner: "x is soluble in water" = df. "Whenever x is put into water, x dissolves." Symbolically this reads as follows: $Sx \equiv (t) (Wxt \supset Dxt)$ where "Sx" stands for "x is soluble in water," "Wxt" for "x is put in water at time t," and "Dxt" for "x dissolves at time t." The difficulty with this formulation, Carnap points out, is that for any object which has never been placed in water the antecedent of the definiens, Wxt, is false and the entire right hand side is thus true; i.e., it follows that any object which was never put in water is therefore soluble in water. The same problem will arise with any attempt to define a disposition term explicitly since the very notion of a disposition will require that the definition be in hypothetical form.

In line with our central theme of exhibiting the role of the logical empiricist's presuppositions in their philosophy of science, there are two points that are worth making here. The first is that, in typical logical empiricist fashion, Carnap's discussion, which is widely regarded by logical empiricists as one of the landmarks in the development of their philosophy of science, deals with relatively simple concepts from everyday experience rather than actual theoretical terms from science, and with formulations that can be conveniently developed in terms of the machinery of principia logic. The assumption that this procedure will somehow clarify the nature of scientific concepts is a fundamental presupposition of the logical empiricist research project and, as such, is neither questioned nor even explicitly stated. Second, and more importantly, the difficulty that Carnap finds with the proposed explicit definition of disposition terms is generated by his acceptance of the principia notion of material implication as a completely adequate tool for formulating scientific propositions.

For, as we have seen, one of the characteristic features of this analysis of implication is the somewhat paradoxical one that any hypothetical proposition with a false antecedent is true and it is this so-called "paradox of material implication" that is responsible for the difficulty Carnap points out. Again, this is not intended as a criticism of Carnap's work, but rather as a further illustration of the role which a philosopher's presuppositional framework plays in generating his problems. It is the empiricist thesis that only terms which are defined by reference to observables have cognitive meaning that led to the attempt to define disposition terms by means of observables, and it is the logical empiricist requirement that such definitions be formulated in the symbolism of principia logic that led to the rejection of the proposed definition and the need for further research.

In order to resolve the above problem Carnap proposed the new method of reduction sentences for introducing disposition terms and thus all theoretical terms into scientific discourse.[12] Consider a new predicate R that we wish to introduce into our language and let P and S denote test conditions that we can bring about, e.g., experimental procedures. Then the following constitute a reduction pair for R:

   (1) $P \supset (Q \supset R)$ and

   (2) $S \supset (T \supset \sim R)$

where Q and T are observables which are possible results of experiments.[13] For example, the first reduction sentence for "soluble" might read, "If an object is placed in water, then if it dissolves, it is soluble." In the special case in which P and S are identical and Q and T are identical we get a single bilateral reduction sentence of the form "$P \supset (Q \equiv R)$." We might, for example, introduce "soluble" by the statement, "If an object is placed in water, then it is soluble if and only if it dissolves." Clearly the introduction of disposition terms by reduction sentences solves the problem that Carnap was originally concerned with. For although the reduction sentence is still true in the case in which P does not occur, it no longer follows that the object in question has the property R, e.g., it no longer follows that any object that has not been put in water is soluble. It does, however, follow that within the limits of the reduction sentences for "soluble" considered above, the term "soluble" is *undefined* for any object which has not been put in water. Reduction sentences introduce terms only for the specified test conditions; they do not provide general definitions of these terms.

Let us consider how Carnap's reduction sentences fit into the empiricist program. It will be recalled that the thesis of that program that we are concerned with here is that every theoretical term must receive its meaning from observation terms. As originally conceived this required explicit definition, which would in principle make it possible to eliminate theoretical terms from scientific discourse, but we have seen that this strong form of the program raises formidable difficulties. Carnap's proposal constitutes a major weakening of this program. Rather than defining theoretical terms by reference to observables, theoretical terms are given a partial interpretation for a particular set of experiments and a particular set of possible outcomes of these experiments. Since these experiments and outcomes (Carnap introduces the terms "realizable" and "observable" to distinguish them[14]) enter the reduction sentences as proper empirical predicates, Carnap may still be viewed as working towards the general empiricist goal. For although he has given up the notion of elimination of theoretical terms in favor of observables, he maintains that terms introduced by reduction sentences are "reduced in a certain sense"[15] to observables.

Reduction sentences only introduce terms for a specific set of specified experimental conditions; the response of empiricists to this approach has been mixed. As might be expected, some of the objections to reduction sentences have come from empiricists who are unwilling to give up the original empiricist notion that to define a term is to show how to eliminate it. In a review of Carnap's paper, for example, Leonard writes: "As a form of postulate, the reduction sentence is certainly unobjectionable, and that it does 'reduce' the predicate in question is undeniable. But that it is appropriate as a device for introducing a new term is certainly dubious."[16] Similarly, Goodman writes: "There are just two ways of introducing terms into a system: (1) as primitives, (2) by definition. Passages in the Carnap article . . . have given rise to the impression that there is a new, third, method of introducing terms: by reduction sentences. . . . This is rather misleading; for to introduce a term by means of reduction postulates is to introduce it as an ineliminable primitive."[17]

Hempel, on the other hand, sees the very incompleteness of meaning of terms that are introduced by reduction sentences as a virtue since it provides a way of accounting for the scientific practice of extending theoretical terms into new domains of experience, a practice which, as we have already seen, cannot be accounted for by what Hempel calls the "narrower thesis of empiricism,"[18] i.e., the thesis that "Any term in the vocabulary of empirical science is definable by means of observation terms. . . ."[19] Thus, discussing the fact that reduction sentences only partly determine the meaning of terms they introduce, Hempel writes:

> It may be well, therefore, to suggest that this . . . characteristic of reduction sentences does justice to what appears to be an important characteristic of the more fruitful technical terms of science; let us call it their *openness of meaning*. The concepts of magnetization, of temperature, of gravitational field, for example, were introduced to serve as crystallization points for the formulation of explanatory and predictive principles. Since the latter are to bear upon phenomena accessible to direct observation, there must be "operational" criteria of application for their constitutive terms, i.e., criteria expressible in terms of observables. Reduction sentences make it possible to formulate such criteria. But precisely in the case of theoretically fruitful concepts, we want to permit, and

indeed count on, the possibility that they may enter into further general priciples, which will connect them with additional variables and will thus provide new criteria of application for them. We would deprive ourselves of these potentialities if we insisted on introducing the technical concepts of science by full definition in terms of observables.[20]

That is, since reduction sentences only define the terms that they introduce for a specified set of circumstances, it remains possible to extend the term to new circumstances by the introduction of new reduction sentences for those new circumstances by the introduction of new reduction sentences for those new circumstances. Consider, for example, a term R which was originally introduced by the reduction pair "$P \supset (Q \supset R)$" and "$S \supset (T \supset \sim R)$." If some new set of regularities involving R is found in some new set of circumstances, we can further specify the meaning of R for these new conditions by introducing a reduction pair, say "$A \supset (B \supset R)$" and "$C \supset (D \supset \sim R)$" and so forth as new situations arise. But this very flexibility also leads to a new set of problems.

For the traditional empiricist all definitions are analytic statements, and from analytic statements we can only deduce further analytic statements. But, as Carnap recognized,[21] if reduction sentences are accepted as a form of definition, the general thesis that all definitions are analytic must be given up, for any reduction pair has a definite empirical content. From the pair "$P \supset (Q \supset R)$" and "$S \supset (T \supset \sim R)$," "$\sim (P \cdot Q \cdot S \cdot T))$" follows, and this is not an analytic statement. Thus the acceptance of reduction sentences as a way of introducing new terms requires that the empiricist give up either the thesis that new terms can only be introduced by analytic statements or the thesis that analytic statements only entail other analytic statements. At the time he wrote *Testability and Meaning* Carnap thought that this problem existed only for the case of reduction pairs and not for bilateral reduction sentences since, in the case of a bilateral reduction sentence such as "$P \supset (Q \equiv R)$," the "content" becomes "$\sim (P \cdot Q \cdot \sim Q)$," which is analytic. Thus Carnap held that "every bilateral reduction sentence is analytic."[22] But Hempel has since pointed out[23] that once we allow the extension of a concept that has been introduced by means of a bilateral reduction sentence into new contexts by introducing further bilateral reduction sentences, the same problem arises. Suppose, for example, that R is defined for two different situations by the bilateral reduction sentences "$P \supset (Q \equiv R)$" and "$A \supset (B \equiv R)$." From these two we can deduce "$(P \cdot Q \cdot A) \supset B$," which is synthetic,[24] and the empiricist is faced with the same choice as in the case of reduction pairs.

The point that must be emphasized for our purposes here is that this difficulty is, once again, generated by the presuppositions of logical empiricist philosophy of science; in particular, in this case, by the demand that all scientific terms must in some way be defined by reference to observables along with the insistence on a strict distinction between analytic and synthetic propositions.

Thus, it is possible for a philosopher to accept reduction sentences as a means of introducing new terms by giving up the analytic/synthetic distinction (Hempel, for one, seems to be willing to at least contemplate this option[25]); or he can maintain the analytic/synthetic distinction and other aspects of the logical empiricist program and surrender both reduction sentences and the attempt to supply empirical content to theoretical terms; or he can simply reject reduction sentences and continue to work toward the fulfillment of the program, since the failure of a particular attempt to carry out a research program does not constitute a refutation of that program (although repeated failures do, of course, supply good reasons for suspecting that a different research program with a different set of presuppositions might be more fruitful). The third alternative, the search for alternative ways of carrying out the program, has of course been the response of most empiricist philosophers of science.

Craig's Theorem

It has been noted that many empiricists take the problem of defining theoretical terms as equivalent to showing how they can be eliminated from scientific discourse. In the approaches we have considered thus far the notion of "elimination" has been taken as meaning "replacement by an equivalent expression," but there is another sense of "elimination" that is relevant to the present problem: a term can also be eliminated from scientific discourse by showing that it is unnecessary, that we can say everything we want to say and do everything we want to do without having to use it. If we can eliminate theoretical terms in this way, the problem of finding their empirical content is dissolved. Now for many empiricists the sole function of science is to find relations between observation statements; theoretical statements (i.e., statements which include theoretical terms) function only as intermediaries in this process. Thus if a way can be found to formulate all connections between observables without having to make use of theoretical terms, theoretical terms will have been shown to be unnecessary and the problem of analyzing their empirical meaning eliminated. One widely discussed attempt to accomplish this goal has developed around a theorem in formal logic proved by William Craig,[26] in spite of Craig's own insistence that "the method is artificial and the solutions it yields are philosophically quite unsatisfactory."[27]

What Craig's theorem provides is a general method of eliminating a selected group of terms from a formalized system without changing the content of the system. In order to apply the method it is first necessary that we have an effective criterion for distinguishing the "essential" expressions of the system from the "auxiliary" expressions. Taking the "content" of the system to be identical with the class of essential expressions, Craig gives a method for constructing a new axiomatized system which includes all the essential expressions and none of the auxiliaries. For a philosopher who views a science as a deductive system cast in terms of principia logic and the essential expressions as those which contain only observation terms, Craig's theorem would seem to offer a way of eliminating theoretical terms. Unfortunately, Craig's method has a number of defects when viewed as an approach to this problem, defects which Craig himself was the first to point out. Nevertheless, a number of empiricist writers have discussed the theorem as a possible way of solving their problem of theoretical terms (and then repeated Craig's own objections).

The most important defect of Craig's method from the point of view of philosophy of science is that it can only be applied to completed deductive systems, i.e., to deductive systems from which we have already drawn all consequences which we will ever draw. Once we have this complete list, Craig gives a method for constructing an infinite and highly redundant axiom set which includes all these consequences among the axioms. Applying this to the relation between theoretical statements and observation statements, the theorem tells us that once we have drawn all observational consequences that can be drawn with the help of our axiomatized theory, we can, in effect, throw out the theory and take these observation statements as (a part of) the axioms for a new deductive system without losing any of the observation statements. But not only does this result eliminate the main purpose of constructing axiomatic systems, i.e., encapsulating a great body of information in a small set of axioms,[28] it also assumes a wholly inadequate picture of science as consisting of unchanging deductive systems,[29] and it fails to show that theoretical terms are unnecessary for science. At most it shows that once we have learned everything we can possibly learn from a theoretical system, then if we are only interested in a certain body of conclusions drawn from this system, we can ignore the system and pay attention only to those conclusions; but it is difficult to see why we needed a theorem of formal logic to tell us that.

Aside from the significance of Craig's theorem within mathematical logic, the most interesting question raised by it is why so many philosophers have found it worthwhile to discuss as a possible solution to the empiricist problem of theoretical terms.[30] This question is best answered by viewing the discussions of Craig's theorem within the context of the logical empiricist research program. Having accepted symbolic logic as the fundamental tool for the analysis of science, it takes any theorem of symbolic logic as possibly throwing light on the nature of science and thus as prima facie worthy of discussion. It is also worth noting that this entire discussion only makes sense from the point of view of a philosopher who assumes that the primary task of science is to find connections between observables. For a different kind of philosopher of science, one who holds, for example, that the goals of science include explaining phenomena or attempting to discover the underlying structure of reality, the construction of theories is a primary aim of the scientific enterprise and the problem of finding a way to eliminate theories would never arise.[31]

There is one more approach to the problem of the empirical significance of theoretical terms which must be examined here, an approach which has more recently gained wide acceptance among contemporary empiricists although it was formulated by Norman Campbell in 1920.[32] Viewing a scientific theory as an axiomatized formal system, a distinction is made between two parts of the formal system: the body of theoretical propositions which is formulated solely in the theoretical vocabulary, and a set of "correspondence rules" which connect functions constructed out of the theoretical terms with observation terms. The distinction can be illustrated by a trivial but clear example adapted from Campbell.[33] Suppose that our theory contains the following theoretical terms:

1) constants a and b,
2) variables c and d.

There is one theoretical claim:

3) $c = d$

And there are two correspondence rules:

4) $(c^2 + d^2)a = R$ where R is the resistance of a metal, an observable,
5) $cd/b = T$ where T is the temperature of the metal, another observable.

As an example of how such a theory might be used note that we can deduce from the theoretical propositions alone the formula "$(c^2 + d^2)a/(cd/b) = 2ab = $ constant." Interpreting this formula in terms of the correspondence rules, it tells us that the ratio of the resistance to the temperature of a metal is a constant and, assuming this to be a law of nature, the theory provides an explanation of this law and could have predicted the law as well if it were not known at the time that the theory was proposed.

What is important about this analysis in the context of our present concerns is that the terms appearing in the theoretical propositions alone, independently of the correspondence rules, are taken as having no empirical meaning; empirical meaning is conferred on these terms when they are connected with experience by means of the correspondence rules. Some empiricist writers allow another kind of meaning which is conferred on theoretical terms solely by virtue of their appearance in the axiom system of the theory. This aspect of their meaning is often discussed under such rubrics as "definition by postulate," "implicit definition" and "systematic import," but the empiricist must hold that this is only a minor part of the meaning of theoretical terms and that the important scientific aspect of their meaning is the empirical meaning that they receive through being connected to observables by correspondence rules. But unlike the previous approaches we have discussed, this approach surrenders the goal of finding a definition in terms of observables for *each* theoretical term. In general no theoretical term will appear alone in a correspondence rule; rather it will appear as part of a function of theoretical terms, and some theoretical terms will not appear in correspondence rules at all. These terms will get their empirical meaning, so to speak, second hand, by virtue of their occurrence in formulas in which other terms which occur in correspondence rules also occur. It is still maintained that it is the correlation with experience that gives meaning to the theoretical terms, but rather than the individual terms receiving meaning from observables, it is the theoretical system as a whole which receives this meaning.

One of the clearest recent explications of this approach is given by Feigl in figure 1:[34]

The theory itself is made up of a set of primitive concepts (i.e., theoretical concepts) which are connected together by postulates. Feigl is willing to concede that the primitives receive some meaning via implicit definition merely from their participation in the system of postulates, but, he writes:

> It is important to realize that implicit definition thus understood is of a purely syntactical character. Concepts thus defined are devoid of empirical content. One may well hesitate to speak of "concepts" here, since strictly speaking even "logical" meaning as understood by Frege and Russell is absent. Any postulate system if taken as (erstwhile) *empirically uninterpreted* merely establishes a network of symbols. The symbols are to be manipulated according to preassigned formation and transformation rules and their "meanings" are, if one can speak of meanings here at all, purely formal.[35]

Other concepts are defined in terms of these primitive concepts and at least some (in Feigl's simplified diagram, all) of these defined concepts are given meaning by correspondence rules

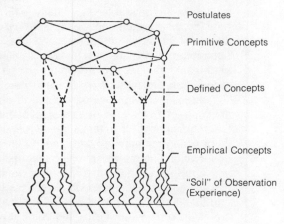

Postulates

Primitive Concepts

Defined Concepts

Empirical Concepts

"Soil" of Observation (Experience)

Figure 1. Feigl's Diagram

which connect them with empirical concepts, the latter being defined directly by reference to experience. (It should be noted that for Feigl it is at this last stage of defining empirical concepts in terms of experience that a form of operational definition comes into play.[36]) The entire structure now becomes meaningful, thanks to "an 'upward seepage' of meaning from the observational terms to the theoretical concepts."[37]

But the notion of correspondence rules too has come under attack from within the empiricist camp. For within the context of the logical empiricist framework, correspondence rules must be taken to be either analytic propositions or rules telling us how to use certain terms, "metalinguistic principles which render certain sentences true by terminological convention or legislation."[38] But, as Hempel goes on to point out, correspondence rules change as a result of empirical research. The scientific concept of time, for example, may be introduced at a given stage in the development of science by taking some periodic process as the standard of equal time intervals. It would seem, then, that the definition of equal time intervals has now been fixed by convention and that no empirical discoveries could force us to change our standard. But, as the history of horology clearly shows, this is not the case: "certain laws or theoretical principles originally based on evidence that includes the readings of standard clocks give rise to the verdict that these clocks do not mark off strictly equal time intervals."[39] Thus correspondence rules are affected by empirical data and they cannot be taken to be either analytic statements or conventional rules. Once again we find that a proposed solution to a philosophical problem generated by the logical empiricist's presuppositional framework fails because it does not meet criteria of adequacy dictated by that framework.

We have come a long way from Russell's formulation of the empiricist program and have seen in some detail how the attempt to carry out this program has brought its continual liberalization and has led logical empiricists progressively to recognize how complex and indirect the connection between observation and high level theoretical terms actually is. Empiricists might reply here that the development we have been examining illustrates the flexibility and open-mindedness of empiricist philosophy, but the open-mindedness appears to have resulted in the transformation of at least this aspect of the empiricist program beyond recognition. By the time we arrive at Feigl's diagram and "seepage" metaphor one cannot help but wonder if there is anything left of the empiricist project of analyzing the meaning of theoretical concepts in terms of experience other than the name. Feigl's diagram may provide a sound illustration of how observation enters into scientific theorizing, but it is only by insisting that scientific theories *must* get their meaning by correlation with observation that it is possible to take the next step and conclude that at whatever point observation enters into scientific knowledge, that is the point at which, somehow, previously meaningless abstract calculi become meaningful theories.

One particularly intriguing result of this development is found in Hempel's recent paper, "On the 'Standard Conception' of Scientific Theories." Hempel, who has been a major contributor throughout this development, now formulates the problem of defining theoretical terms as one of defining them in terms of an antecedently available vocabulary where this vocabulary is, in general, not observational in that it includes terms which were introduced in the context of earlier theories.[40] Hempel's new formulation of the problem amounts to a rejection of the sharp distinction between observation and theory which generated the problem in the first place,[41] and his discussion of this new, even more liberalized version of the problem, leads him to conclude, following Putnam, that the "presumptive problem 'does not exist.' "[42] Hempel now maintains that the important problem is not how we define new theoretical terms but how we come to *understand* them and he rejects the thesis that we do this only by specifying their meaning in terms of previously understood terms.

> We come to understand new terms, we learn how to use them properly, in many ways besides definition: from instances of their use in particular contexts, from paraphrases that can make no claim to being definitions, and so forth.[43]

What Hempel fails to note is that this new position does not constitute an abandonment of the general problem of the meaning of theoretical terms, but rather, as we shall see in Part II, a statement of a different, non-empiricist, approach to the solution of this problem.

Explanation

It is generally agreed among logical empiricists that the basic pattern of scientific explanation is the deductive pattern and that this applies equally to the three major areas of scientific explanation: explanation of events by means of laws, of laws by means of theories, and of theories by means of wider theories.[1] Once again there is a classic statement of this view around which much of the subsequent discussion has centered, "Studies in the Logic of Explanation" by Hempel and Oppenheim.[2] The authors propose four conditions of adequacy for a scientific explanation.[3] Three of these are labelled "logical conditions": (1) The explanadum must be logically entailed by the explanans; (2) the explanans must contain general laws which are necessary for the deduction of the explanandum (in the case of explanation of an event, the explanans must also contain statements of antecedent conditions, i.e., statements which refer to specific empirical objects or events); (3) the explanans must have empirical content. The fourth condition, referred to as an "empirical condition," is that the explanans must be true, not just well confirmed.[4] The deductive model is summed up in the following diagram:[5]

$$
\text{Logical Deduction}
\left\{
\begin{array}{l}
\left.
\begin{array}{ll}
C_1, C_2, \ldots C_k & \text{antecedent conditions} \\
L^1{}_1, L_2, \ldots L_r & \text{General Laws}
\end{array}
\right\} \text{Explanans} \\[1em]
\left.
\begin{array}{l}
E \quad \text{Description of empirical} \\
\quad \text{phenomenon to be explained}
\end{array}
\right\} \text{Explanandum}
\end{array}
\right.
$$

It should not be surprising at this stage of our discussion to find that Hempel and Oppenheim formulate the problem of analyzing scientific explanation as a logical problem, and that a significant part of their paper (as well as of the subsequent literature) is concerned with generating logical puzzles which result from the conditions of adequacy and with seeking means to amend those conditions to remove the puzzles. For example, Hempel and Oppenheim themselves point out that in the case of explanations of particular events their conditions of adequacy allow any general law to explain any event.[6] Let E, the event to be explained, be "Mount Everest is snowcapped," and let L be a general law such as "All metals are good conductors of heat." We can now construct the following statement of initial conditions: let L' be a specific instance of L such as "If the Eiffel Tower is metal, then it is a good conductor of heat," and let C, the statement of initial conditions, be "$L' \supset E$." C is true since E is true and the conjunction of L and C now constitute, according to the criteria of adequacy, an explanation of E. This, it will be recognized, is an anomalous result.

Hempel and Oppenheim deal with this anomaly by formulating an additional logical requirement which will remove it. First we define a *basic sentence* as either an atomic sentence or the negation of an atomic sentence and the *verification of a molecular sentence* as deduction from some class of true basic sentences. The additional requirement proposed is that there must be a class of basic sentences from which we can deduce C but neither $\sim L$ nor E.[7] That this eliminates the problem can be seen from the following considerations: the statement of initial conditions that was constructed, "$L' \supset E$," is itself logically equivalent to "$\sim L' \vee E$" and this is true if either of two conditions is met. It is true if L' is false, but this would entail that L is false and we know, by hypothesis, that this is not the case. Or it is true if E is true, and we know that this is the case since E is the event to be explained. The proposed additional criterion stipulates that we must have some means of verifying C which is logically independent of means of verifying either L or E and thus eliminates the possibility of establishing the truth of C solely on the basis of information about L and E.[8] But there is more involved here than appears at first glance.

To begin with let us note the role of the logical part of the logical empiricist framework in generating this problem. It is in accordance with this framework that Hempel and Oppenheim seek a logical analysis of explanation, and in particular one which can be formulated in the symbolism of principia logic, but the problem we have been discussing is a result of their use of material implication in formulating this analysis. For while they have given a completely general method for constructing propositions such as C for a given L and E, the conditions of adequacy require that C be true and the truth of C in their example is guaranteed by the truth of E (which is known to be true since it is the explanandum). The entire difficulty is thus generated by one of the paradoxes of material .

implication: *any* hypothetical statement with a true consequent is true irrespective of whether the antecedent and consequent have anything to do with each other. It will require a somewhat more complex analysis to locate the role of the empiricist aspect of the presuppositional framework in this discussion.

In a widely discussed passage Hempel and Oppenheim pointed out that explanation and prediction have the same logical form.

> Let us note here that the same formal analysis, including the four necessary conditions, applies to scientific prediction as well as to explanation. The difference between the two is of a pragmatic character. If E is given, i.e. if we know that the phenomenon described by E has occurred, and a suitable set of statements C_1, C_2, . . . , C_k, L_1, L_2, . . . , L_r is provided afterwards, we speak of an explanation of the phenomenon in question. If the latter statements are given and E is derived prior to the occurrence of the phenomenon it describes, we speak of a prediction. It may be said, therefore, that an explanation of a particular event is not fully adequate unless its explanans, if taken account of in time, could have served as a basis for predicting the event in question. Consequently, whatever will be said in the article concerning the logical characteristics of explanation or prediction will be applicable to either, even if only one of them should be mentioned.[9]

It is in accordance with this point that Hempel and Oppenheim, diagnosing the source of the problem we have been considering, write, "the peculiarity just pointed out clearly deprives the proposed potential explanation for [E] of the predictive import which . . . is essential for scientific explanation. . . ."[10] That is, as we have seen, we can only establish the truth of C because we already know E to be true, and this eliminates the possibility of having used C as a premise in an argument which could predict E. Indeed, the solution to the problem which Hempel and Oppenheim propose consists of adding the requirement that we must be able to establish the truth of C independently of the truth of E. This guarantees that L and C will allow us to predict E, and only then can it be said that the conjunction of L and C explain E.

Put somewhat differently, the defect in the method that Hempel and Oppenheim construct for using any law to explain any event is seen by them to be equivalent to eliminating the predictive force of the proposed explanation and the anomaly is resolved by restoring this predictive force. But from this point of view, Hempel and Oppenheim's statement that explanation and prediction have the same "formal structure" is somewhat misleading. This notion is originally introduced as an observation on the conditions of adequacy, but Hempel and Oppenheim themselves offer a counter-instance to this claim, a case of a formally satisfactory explanation which is non-predictive, and they restore the generality of the prediction-explanation parallel by introducing an additional condition of adequacy which says, in effect, that E is only explained from a set of premises if it could have been predicted from those premises. Thus, in maintaining the structural identity of explanation and prediction, Hempel and Oppenheim did more than make an interesting logical observation: prediction serves as a desideratum for any adequate explanation and in fact as a fifth criterion of adequacy. And we should note that this is at least hinted at in their original formulation of the structural identity thesis when they write, "It may be said, therefore, that an explanation of a particular event is not fully adequate unless its explanans, if taken account of in time, could have served as a basis for predicting the event in question,"[11] without adding, "and conversely." Indeed, the converse of this statement, "A prediction of a particular event is not fully adequate unless its premises could serve as an explanation of that event," would be somewhat odd, for how could a correct prediction fail to be a fully adequate prediction?

Considerable light can be thrown on the role that the prediction criterion plays in the above analysis if we look at it from another point of view. It has long been recognized that there are an unlimited number of sets of premises from which any given statement can be deduced, and probably even an unlimited number of sets of true premises—particularly in the context of principia logic, where the paradoxes of material implication make it rather easy to construct true hypothetical statements. A philosopher who advocates that explanation consists of deduction from true[12] premises faces, then, a problem of supplying some additional criterion for distinguishing explanatory deductions from non-explanatory deductions. One important approach to this problem is to maintain that besides being formally satisfactory, an acceptable explanation must also provide a familiar model or an analogy with familiar situations.[13] But Hempel rejects this thesis as being irrelevant to his problem, which is to analyze the *logic* of explanation. Discussing Campbell's analysis of the role of analogy in scientific explanation, Hempel writes, "Campbell fails to establish that analogy plays an essential logic-systematic role in scientific theorizing; some of his pronouncements squarely place his requirement of analogy within the domain of the pragmatic-psychological aspects of explanation."[14] In general, Hempel maintains, "For the systematic purposes of scientific explanation, reliance on analogies is thus inessential and can always be dispensed with."[15] For Hempel, models and analogies may play an important heuristic role in the process of constructing theories and a pragmatic role in helping us to understand theories,[16] but they have no relevance to studies of the logical structure of explanation. We come up, once again, against the research project that Hempel is engaged in. To accept models or analogies as a part of explanation is to abandon his presuppositional base, and at least one contemporary advocate of a models approach to explanation has recognized this in claiming that his approach constitutes a Copernican revolution in the philosophy of science.[17]

But Hempel and Oppenheim still must deal with the problem of how they distinguish arguments which are both deductive and explanatory from those which are deductive without being explanatory. I submit that the criterion used is prediction, that only deductions from those premises from which the phenomenon in question could have been predicted constitute adequate explanations.[18] Indeed, in a more recent essay, in response to continuing attacks on the thesis that explanation and prediction are logically indistinguishable, Hempel has abandoned this thesis in its original form. He now holds that not all predictions are explanations, but he continues to hold that all explanatory arguments are predictive.[19] Prediction is, then, the wider concept and explanations are a sub-class of predictions. Thus the controlling factor in Hempel's discussions of the relation between explanation and prediction is the primacy of prediction and the problem of distinguishing explanatory deductions from mere deductions is, in effect, resolved by invoking the demand that explanations must be predictive. This is completely in keeping with the strong strain in empiricist philosophy which maintains that the primary function of scientific knowledge is the prediction of observables.

As has been noted above, the thesis of the structural identity of explanation and prediction has been widely attacked. It will be useful for our purposes here to examine one of the most fully developed and persistent of these attacks, that of Michael Scriven. Scriven has constructed a number of proposed counter-instances to the thesis that explanation and prediction are equivalent, counter-instances intended to show that we can often explain events which we could not have predicted. The following example is typical: We can explain a patient's having paresis by reference to his having syphilis since syphilis is the only cause of paresis. But since only a small percentage of syphilitic patients develop paresis, we could not have predicted that this patient would develop paresis.[20] To this line of argument Hempel responds,

> Precisely because paresis is such a rare sequel of syphilis, prior syphilitic infection surely cannot by itself provide an adequate explanation for it. A condition that is nomically necessary for the occurrence of an event does not, in general, explain it; or else we would be able to explain a man's winning the first prize in the Irish sweepstakes by pointing out that he had previously bought a ticket, and that only a person who owns a ticket can win the first prize.[21]

But Scriven's counter-instance is only a small part of a much more complex argument which Hempel ignores and, indeed, Scriven's rejection of the equivalence of explanation and prediction is only a consequence of a much more fundamental disagreement with Hempel.

Scriven's attack is directed primarily at the deductive model of explanation. Two assumptions are fundamental to Scriven's approach: First, that scientific explanation is only a special case of "ordinary" explanation, so that an adequate analysis of explanation requires us to direct our attention to all forms of discourse that are referred to as explanations in everyday discourse. Second, that the function of explanations is to produce understanding, and thus that any instance of discourse which provides understanding is an explanation no matter what its formal structure. It is in this vein that Scriven writes:

> It seems reasonable to suppose that the scientific explanation represents a refinement on, rather than a totally different kind of entity from, the ordinary explanation. In our terms, it is the *understanding* which is the essential part of an explanation. . . . We shall argue that good inductive inferribility is the only required relation involved in explanations, deduction being a dispensable and overrestrictive requirement which may of course sometimes be met.[22]

Similarly, further on in the same essay, Scriven writes:

> Who is to say whether S and Y are understood? The *primary* case of explanation is the case of explaining X *to someone;* if there were no cases of this kind, there could be no such thing as "an explanation of X" in the abstract, whereas the reverse is not true. For it makes no sense to talk of *an explanation* which nobody understands now, or has understood, or will, i.e., which is not *an explanation for someone.*[23]

In effect, Scriven maintains that explaining is a form of discourse which takes place in a specific context among some group of speakers and that instead of analyzing the explanandum and the explanans as Hempel and Oppenheim do, we should, perhaps, examine the explainer and the explainee. The explainer's task is to get the explainee to understand something; anything that the former can say or do which will accomplish this purpose counts as an explanation.

But while Scriven's analysis may indeed be an accurate description of the way in which we use the term "explain" when we are engaged in such everyday activities as explaining our symptoms to a doctor or explaining how a lever works to a child, it is not at all clear how this is relevant to the work of Hempel and Oppenheim. They were not attempting to analyze ordinary language uses of "explain," but to provide a logical reconstruction of scientific explanation, and this problem, as we have seen, is one which is dictated by the philosophical framework from which they approach the philosophy of science. From that point of view, attempts to provide an analysis of ordinary language are irrelevant. To a large extent, then, the argument between Hempel and Scriven is at cross-purposes, since they have different conceptions of what the problem of explanation is, and thus they are not trying to solve the same problem at all. What Scriven proposes is not an alternative analysis of explanation within the logical empiricist framework, but rather the reformulation of the problem of explanation from the viewpoint of a different philosophical framework (what is commonly referred to as "ordinary language philosophy"). The level on which Hempel and Scriven do disagree is on the question of which research project is more worth pursuing as an approach to the philosophy of science. In so far as Hempel declines Scriven's invitation to abandon his research project and adopt another, his ignoring of all of Scriven's analysis and argument except the proposed counter-instances is appropriate. Indeed, the manner in which Hempel responds to Scriven's counter-instances, by asserting that they are not adequate explanations since they do not demonstrate a necessary (i.e., deductive) tie between explanans and explanandum, is a good index of the degree to which the arguments are at cross-purposes: Scriven proposing what he takes to be a clear case of non-deductive explanation and Hempel responding that it is not an adequate example of an explanation since it is non-deductive.

My claim that the debate between Scriven and the deductivists is at cross-purposes receives further support from an attempt by Brodbeck to provide a detailed response to Scriven from the deductivist point of view.[24] For while Hempel ignores Scriven's arguments, Brodbeck's responses often come to no more than mild sarcasm and expressions of astonishment. One example will suffice:

> The only way Scriven persuades himself that he can explain an event that could not even in principle be predicted is by leaving "causal" statements wholly unanalyzed. Despite the confident use of the causal idiom in everyday speech, we may still significantly ask under what conditions statements like "C is the cause of E" are true or false. I shall not take the time here to exhibit the problematic nature of the notion of "cause." Nor do I believe that to most this needs exhibiting.[25]

One of the most striking characteristics of debates between thinkers who are working from different presuppositional bases is that along with disagreements on what problems need to be solved and what constitute adequate solutions of these problems, they also disagree on which concepts do and which do not require explication and find it unnecessary (if not impossible) to offer arguments for their choice.

By way of comparison, and to reinforce the point that not all philosophical disagreements result from the acceptance of different presuppositional frameworks, but that genuine disagreements are possible among writers who accept the same framework, let us consider an objection to the thesis of the equivalence of prediction and explanation that has been raised by Scheffler. Scheffler is himself a strict deductivist. In a discussion of statistical explanation, for example, he holds that to explain why Jones had a heart attack by pointing out that 75 percent of the men in his age group have heart attacks leaves out exactly what we wish to explain, i.e., why *Jones* had a heart attack.[26] To answer this question we must, according to Scheffler, collect other information about Jones, information that would supply sufficient additional premises to permit us to deduce that Jones had a heart attack. Scheffler recognizes that statistical explanations of the above sort are used in science, describes them as "a special case of pragmatically incomplete deductive explanation,"[27] and concludes that it seems reasonable to extend the concept of explanation to include statistical explanation; but note that Scheffler considers this an *extension* of the concept. Now Scheffler also rejects the thesis that explanations and predictions are structurally identical, holding that not all predictions are deductive and that we can thus often predict events which we could not explain. An observed constant conjunction of A and B, for example, might lead us to predict B on observing A, but it would supply no grounds for an explanation of B.[28] This position is perfectly compatible with the logical empiricist framework, it is one which Hempel now accepts, and it is also an objection of a completely different order from Scriven's objections: Scheffler is attempting to clarify a point within the structure of the logical empiricist framework, while Scriven is attempting to overturn that framework.[29]

Statistical Explanation

We turn now to an examination of the second major kind of scientific explanation recognized by logical empiricists, statistical explanation. Unlike a deductive explanation, a statistical explanation does not show that given the premises, the phenomenon to be explained necessarily occurs, but only that it is highly probable, or, perhaps, almost certain. As a fairly simple example of a statistical explanation consider the following: Jones has a streptococcal infection and recovers after being treated with penicillin. Since we know that almost all of those persons who have streptococcal infections recover when treated with penicillin, we could propose the following explanation:

The probability of recovering from a streptococcal infection when treated with penicillin is close to one.

Jones had a streptococcal infection and was treated with penicillin.

———————————————————————————————
———————————————————————————————

Jones recovers from the streptococcal infection.

The double line, taken over from Hempel,[30] is used to distinguish the statistical schema from the deductive and should be read as, "Makes highly probable," or "Makes practically certain." But as Hempel, among others, has pointed out, the presentation of statistical explanations in this quasi-deductive form leads to what appears to be an inconsistency. For suppose that we also know that Jones is over 80 years old and that almost all men over 80 years old who develop streptococcal infections do not recover. If Jones failed to recover, we could then offer the following explanation:

The probability of a man over 80 years old not recovering from a streptococcal infection is close to one.

Jones is over 80 years old and has a streptococcal infection.

Jones does not recover from the streptococcal infection.

This result is problematic in two respects. First, it provides us with a case of two arguments which are presumably valid, which have true premises, and which are logically inconsistent with each other: something which can never occur in deductive logic. Second, it shows that we can offer a valid statistical explanation no matter what the outcome of Jones' illness is, which Hempel takes to be a strong indication that something is wrong with the explanation.[31]

Hempel argues that the existence of this *ambiguity of statistical explanation* shows that statistical explanations are of a fundamentally different kind from deductive explanations. Thus at one point he suggests that statistical explanation involves a different sense of the word "because" than does deductive explanation.[32] He also suggests that the ambiguities arise from the attempt to construe statistical explanations as, in some broad sense, syllogistic, and that this construal "seems to aim at too close a formal assimilation of nondeductive statistical arguments to deductive inference."[33] But in spite of these statements, Hempel's actual practice in attempting to resolve the problem of the ambiguities clearly shows that he continues to take deductive explanation as his model and attempts to assimilate statistical explanation as closely as possible to deductive explanation, he does take the ambiguity of statistical explanation as a problem to be resolved. It is thus reasonable to ask why he considers this to be a problem. If statistical explanation is a genuinely different kind of explanation from deductive explanation with genuinely different properties, why not accept the ambiguity as one of these distinguishing properties? Why should we view it as problematic that two different explanations, i.e., two explanations with different (albeit true) premises, can explain two logically incompatible events? Hempel's answer is clear: this never occurs in the case of deductive explanation.[34] Thus deductive explanation still serves as the paradigm case against which all forms of explanation are to be measured.[35] This analysis derives further support from the way in which Hempel resolves the problem of the ambiguity of statistical explanation.

Hempel resolves the problem by invoking Carnap's requirement of total evidence: "In the application of inductive logic to a given knowledge situation, the total evidence available must be taken as the basis for determining the degree of confirmation."[36] This requirement resolves the problem by eliminating all but one of the proposed explanations. In the illustration we considered above, the first explanation, which neglected to include the data on Jones' age and on the probability of an octogenarian recovering from a streptococcal infection, is unacceptable since it violates the requirement of total evidence. And, in general, once the criterion of total evidence is taken into account, only one of a set of mutually inconsistent phenomena can be explained. The upshot of Hempel's discussion of the ambiguity of statistical explanation is the elimination of the ambiguity, in spite of his assertion that "the logic of statistical systematization differs fundamentally from that of deductive-nomological systematization"[37] and the ambiguity of statistical systematization is "one striking symptom of the difference."[38] It is because statistical explanation appears to be different from deductive explanation on this point that statistical explanation is seen as problematic, and it is the demonstration that this difference is only apparent that constitutes, for Hempel, an acceptable resolution of the problem. Thus both the problem of the ambiguity of statistical explanation and the criterion for an acceptable solution of this problem derive from the paradigmatic role of deductive logic in the logical empiricists' research program.

Explanation and Truth

There is one more aspect of scientific explanation that must be considered here, the process by which accepted laws are explained by theories and narrow theories by wider theories. From the point of view of the logic of explanation alone no new issues are raised, but new issues are raised if we set this question in the context of the history of science, for it is here that we find the customary picture of science as a process of cumulative growth of knowledge. According to this picture, theories (and laws) are proposed to cover a particular range of phenomena and are later explained by being subsumed under wider, more comprehensive theories. Nagel, for example, describes this as

> the normal expansion of some body of theory, initially proposed for a certain extensive domain of phenomena, so that laws which previously may have been found to hold in a narrow sector of that domain, or in some other domain homogeneous in a readily identifiable sense with the first, are shown to be derivable from that theory when suitably specialized. For example, Galileo's *Two New Sciences* was a contribution to the physics of freely falling terrestrial bodies; but when Newton showed that his own general theory of mechanics and gravitation, when supplemented by appropriate boundary conditions, entailed Galileo's laws, the latter were incorporated into the Newtonian theory as a special case.[39]

Other stock examples of this process are the explanation of Kepler's laws of planetary motion by Newton's theory and the thesis that Newtonian mechanics is itself explained as a special case of the theory of relativity.[40]

But a moment's reflection will show that the above description of the relation between Galileo's law and Newton's theory is not at all accurate; Galileo's law of falling bodies cannot be deduced from Newton's mechanics, nor even from Newton's mechanics in conjunction with the additional premise that the falling body is "near" the earth. According to Galileo, a body falls to the earth with a constant acceleration, but according to Newton the acceleration is inversely proportional to the square of the distance from the center of the earth and thus increases steadily as the body falls. The fact that for a body sufficiently close to the earth's surface the change in the acceleration is small is quite irrelevant to the point in question, i.e., that Newton's theory, even when supplemented by appropriate boundary conditions, still does not *entail* Galileo's law.

Essentially the same point holds for the relation between Kepler's laws and Newton's theory. According to Kepler, planets have elliptical orbits, and it is true that the application of Newton's laws to a system consisting of any single planet and the sun allows us to deduce that the orbit of the planet is elliptical. But part of what Newtonian theory tells us is that we cannot limit our consideration to just a single planet and the sun, since the other planets also exercise a gravitational attraction on the planet in question, and the resulting orbit is not elliptical. Indeed, the fact that the planets attract each other is an important part of the *universality* of Newton's principle of universal gravitation, which constitutes a major advance over Kepler, who believed that the sun was a fundamentally different kind of body from the planets and that it was the sun alone which moved the planets. Thus, while it is true that Newton's mechanics gives an elliptical orbit in the case of the two body problem, it does not yield an elliptical orbit for the actual case of any of the planets. To ignore the presence of the other planets is not to apply appropriate boundary conditions, but rather to ignore a fundamental part of the content of Newton's theory. It is, of course, perfectly legitimate for the working scientist to make this simplification in doing computations, but it is not legitimate for the logician to make the same simplification when he wishes to maintain that Newton's theory *entails* Kepler's law.

The same point can be made even more strongly in the case of Kepler's third law, that the cube of the mean distance of a planet from the sun divided by the square of the period of revolution is a constant for all of the planets ($a^3/T^2 = $ constant). The law that can be deduced from Newton's theory is that $a^3/T^2 = K(M+m)$ where "K" is a constant, "M" is the mass of the sun and "m" is the mass of the planet in question. But "M+m" is not a constant, since different planets have different masses. It is only by ignoring "m" on the grounds that for the members of our solar system it is much smaller than "M" that we can get a law that looks like Kepler's law. But not only can this law not be deduced from Newton's theory, to ignore "m" consistently is impossible if we are to use Newton's celestial mechanics, since there can be no gravitational force on a body of zero mass ($F = GmM/r^2$).[41] Whatever relation may hold between Newton's theory and Kepler's laws, it is clearly not one of entailment.

A somewhat different appraisal of the situation is given by

Hempel. Discussing the relation between Galileo's law and Newton's mechanics, he writes:

> But while, strictly speaking, Newton's law contradicts Galileo's, it shows that the latter is almost exactly satisfied in free fall over short distances. In slightly greater detail, we might say that the Newtonian theory of gravitation and of motion implies its own laws concerning free fall under various circumstances. According to one of these, the acceleration of a small object falling freely towards a homogeneous spherical body varies inversely as the square of its distance from the center of the sphere, and thus increases in the course of the fall; and the uniformity expressed by this law is explained in a strictly deductive sense by the Newtonian theory. But when conjoined with the assumption that the earth is a homogeneous sphere of specified mass and radius, the law in question implies that for free fall over short distances near the surface of the earth, Galileo's law holds to a high degree of approximation; in this sense, the theory may be said to provide an *approximative D-N explanation* of Galileo's law.[42]

This passage is, in many respects, a strange one. On the one hand Hempel concedes that Galileo's law cannot, strictly speaking, be deduced from Newton's mechanics, yet he still wishes to maintain that the law is explained by that theory; but it is not at all clear what kind of explanation is involved. Hempel appears to be introducing a new category of explanation, approximative deductive-nomological explanations, but no analysis of this kind of explanation is offered. All that we can infer from the name that he gives to this new form of explanation is that it is in some sense a form of deductive explanation, yet it is not clear what forms of deduction are to be allowed other than strict deduction. And it would not be sufficient to reply that all that is meant is that Newton's theory entails a law which gives numerical results that are approximately equal to Galileo's in a particular range of cases. For while this is undoubtedly true, the same can be said of an infinite number of other possible laws, none of which Hempel would describe as explained by Newton's theory.

Similarly, discussing the relation between Kepler's and Newton's laws, Hempel writes:

> In the case of the explanation of Kepler's laws by means of the law of gravitation and the laws of mechanics, the deduction yields a conclusion of which the generalization to be explained is only an approximation. Then the explanatory principles not only show why a presumptive general law holds, at least in approximation, but also provide an explanation for the deviations.[43]

Again we must ask what kind of explanation is involved here. While there may be an intuitive sense in which we would say that Newton's laws explain why Kepler's laws give approximately the correct result, the sense of "explanation" in question here is not the deductive sense. No statement of why Kepler's laws deviate from Newton's can be *deduced* from Newtonian mechanics. Nor does this unanalyzed sense of "explanation" seem to fit any of the other forms of explanation that Hempel has examined. Clearly something has gone awry. It will be instructive to attempt to clarify what has happened.

The crucial feature of the discussion is that in attempting to show how Galileo's and Kepler's laws have been explained by Newton's mechanics, Hempel and Nagel in effect deny that Galileo's and Kepler's laws have been superseded, that they have been shown to be false. For both Hempel and Nagel, the work of Galileo and Kepler are permanent achievements of the scientific method and, as such, cannot be overturned. There is no doubt that, being good empiricists, Hempel and Nagel, along with most other logical empiricists, would reject this claim. All scientific propositions, we are continually reminded, are based on experience and can be overturned by further experience: this is a fundamental thesis of empiricism. Yet in spite of the frequent repetition of this slogan by logical empiricists, an examination of their philosophical practice shows that there is a clear tension between this empiricist principle and the competing belief that, in spite of the fact that the scientific method is empirical, its results are true and stand forever. The latter notion was an explicit part of the logical positivist program, in particular the verification theory of meaning, but the general thesis that a correct application of the "empirical method" will establish scientific truth once and for all goes back at least as far as Bacon. We have seen that contemporary logical empiricists have considerably liberalized the original positivist theory of meaning but, in spite of this, the view that science does provide final truths has remained a controlling presupposition of logical empiricist philosophy of science, while the empiricist disclaimer has been uttered on ritual occasions and used as a weapon against opponents. Let me attempt to document this claim more fully.[44]

The degree to which logical empiricists have vacillated on this question is well illustrated by Reichenbach. On the one hand he takes his fellow empiricists to task for not having adequately appreciated the probable nature of all empirical knowledge:

> The idea that knowledge is an approximative system which will never become "true" has been acknowledged by almost all writers of the empiricist group; but never have the logical consequences of this idea been sufficiently realized. The approximative character of science has been considered as a necessary evil, unavoidable for all practical knowledge, but not to be counted among the essential features of knowledge; the probability element in science was taken as a provisional feature, appearing in scientific investigation as long as it is on the path of discovery but disappearing in knowledge as a definitive system.[45]

But a bit later, discussing whether all thought requires language, Reichenbach writes, "This is a question which psychologists have not *yet* brought to a definite solution."[46] And in *The Rise of Scientific Philosophy* Reichenbach first emphasizes that "Physical theories give an account of the observational knowledge of their time; they cannot claim to be eternal truths,"[47] but then goes on to state that the wave particle duality is an "inescapable consequence of the structural nature of matter,"[48] and that as a result of the experiments of Davisson and Germer "the existence of matter waves was insured beyond doubt."[49] It could, of course, be replied that these are only minor stylistic slips, but a deeper analysis of some central logical empiricist positions will show that this is not the case at all.

We have seen that in "Studies in the Logic of Explanation" Hempel and Oppenheim maintain that "the sentences constituting the explanans must be true."[50] They go on to emphasize that it is not sufficient that the explanans be highly confirmed. For, according to the principle of empiricism, any highly confirmed proposition can be overturned, which would lead us to a situation in which we must hold that an explanation that was once adequate is no longer adequate. But in such a case it is preferable, Hempel and Oppenheim maintain, to reject the former explanation as never having been a genuine explanation.[51] Thus they have made a distinction between propositions that are true and those that are only highly confirmed, and since they apparently held that science does explain, they must also have held that science can get beyond a high degree of confirmation and discover premises which are true, i.e., which can never be overturned. If, by this analysis, Galileo's law, for example, does provide an explanation of free fall near the surface of the earth, then Galileo's law must be true and it cannot be overturned by Newton's theory, even if Newton's theory appears inconsistent with it. The task of the philosopher of science working within the logical empiricist tradition becomes, then, to attempt to find an interpretation of this inconsistency which will allow him to hold that both Newton's and Galileo's laws are true. This is what Hempel attempts to do when he introduces the notion of an approximate deductive explanation.

More recently Hempel has modified his position and recognized that legitimate explanations are possible with premises that are only highly confirmed. This yields two different kinds of explanations: true explanations and explanations that are more or less well confirmed.[52] But there is nothing in Hempel's discussion of this distinction which gives us reason to believe that he has given up the notion that true explanations are achievable in science and the consequent notion that science is capable of establishing the final truth of propositions.

Turning now to Nagel's discussion of this issue, we find that his position is considerably more subtle than that of Hempel. Discussing the epistemic requirements for a satisfactory explanation, Nagel maintains that "the requirement that the premises in a satisfactory explanation must be true seems inescapable,"[53] but at the same time he denies that these premises must be *known* to be true. In support of this latter position Nagel argues that:

In point of fact, we do *not* know whether the unrestrictedly universal premises assumed in the explanations of the empirical sciences are indeed true; and, were the requirement adopted, most of the widely accepted explanations in current science would have to be rejected as unsatisfactory. This is in effect a *reductio ad absurdum* of the requirement.[54]

It would be difficult to find a clearer expression of the tension between the two theses, that all proposed scientific laws are refutab e hypotheses and that science does discover final truths, than the one given in the two passages just quoted. For in the second passage Nagel maintains that any view which would imply that most of the currently accepted scientific explanations are unacceptable is absurd. (Note that Nagel says that *most*, not all, currently accepted scientific explanations would fall if we demanded that the premises must be known to be true. I take it, then, that according to Nagel there are some currently accepted scientific explanations in which the premises are both true and known to be true.) Thus Nagel is maintaining that current science includes a fair number of acceptable scientific explanations. But he has already argued that an acceptable scientific explanation must have true premises, so that he must hold that a fair number of contemporary scientific theories are indeed true. Assuming that Nagel takes the premises of his own *reductio ad absurdum* argument to be both true and known true (and he must do this if he wishes to assert the conclusion), it follows that he holds that we do know that many currently accepted scientific explanations are satisfactory and thus have true premises, but that we do not know which scientific claims fall into this class. Nagel does not explain how we know that the statement, "Many currently accepted scientific claims are true," is true, and in particular he does not explain whether this claim has been established by scientific research or by some other means, but the important point for our purposes is that Nagel does clearly maintain that science has already established a large number of true propositions.

One more example will serve to round out this discussion. In a recent essay Feigl, objecting to those recent writers who maintain that science is based on presuppositions, writes:

> As Reichenbach pointed out a long time ago, science progresses by successively "securing" its various knowledge claims. For example, the optics of telescopes, microscopes, spectroscopes, interferometers, etc., is indeed presupposed in the testing of astrophysical, biological, etc., hypotheses. But these presuppositions—while, of course, always in principle open to (in rare cases actually in need of) revision—are comparatively so much better established than the "farther out" hypotheses that are under scrutiny.[55]

In this passage Feigl pays his ritual obeisance to the in-principle-possibility of overturning "established" scientific knowledge claims, but that it is no more than ritual obeisance is clear. In spite of Feigl's insistence in this essay that the proper domain of the philosopher of science is problems of logic and methodology, Feigl does not consider the logical possibility of overturning accepted scientific knowledge claims to be really important. I submit, once again, that this is because Feigl is operating within the confines of the logical empiricist presupposition that, with the possible exception of "rare cases," science does indeed succeed in securing the final truth of its knowledge claims.

There are two aspects of this discussion that are particularly important for our purposes here. The first is the picture of the history of science that is implicit in the view that science does establish true propositions. According to this view, the history of science, since its inception, has been one of steady accumulation of true propositions. Further research increases our fund of true propositions, and since some of these added propositions are more general than others already established, the unity of science is continually increased. Any scientific proposition is, we are told, in principle refutable, but as a matter of fact, it is only very rarely that further research results in previous scientific accomplishments being overturned. Second, along with this view of the history of science comes a research project for the philosopher who turns his attention to the history of science, since, at least *prima facie*, the history of science does not appear to be one of steady accumulation. (Feigl himself admits that logical empiricists have been remiss in not looking at the actual history of science, but rather making it up a priori!).[56] There appear to be a large number of theories that have served for a while and then been abandoned, many situations in which scientists have continued to make use of theories which they had every reason to believe false, instances in which theories which were believed to be refuted once and for all have reappeared, and even instances in which scientists have refused to accept data which was inconsistent with accepted theories. But, as with any other research project, the logical empiricist need not accept these situations at face value; he can, instead, seek ways to reconcile them with his presupposition as to the nature of the history of science. Among the methods used have been the denial that certain rejected theories, such as phlogiston chemistry and Aristotelian dynamics, were ever science in the first place, and the kinds of efforts that we have seen from Hempel in his attempt to save the truth of Galileo's and Kepler's laws. But a further discussion of this theme is better left for Part II. For the present the important point is to recognize the role of the thesis that science does establish true propositions as one of the guiding presuppositions of logical empiricist philosophy of science.

Chapter Five

Falsification

Before concluding this first part of our discussion, let us examine an approach to the philosophy of science which is, in many ways, transitional between logical empiricism and the new image of science that we will consider in part II. This approach, which has come to be known as *falsificationism,* was introduced by Karl Popper in his *Logic of Scientific Discovery.*[1] Popper's central thesis is that there is no process of induction by which scientific theories are confirmed and thus no role in the philosophy of science for a theory of confirmation as understood by logical empiricists. This would seem to constitute a fundamental break with the research program of logical empiricism, but we shall see that a great deal of Popper's work is nonetheless controlled by just those philosophical presuppositions that we have been examining. Indeed, there are two conflicting strains in Popper's thought. One of these strains is a strict falsificationist view of science, according to which we test scientific theories by deducing consequences from them and rejecting those theories which entail a single false consequence. It is this view of science that has, until recently, generally been attributed to Popper by most philosophers of science,[2] including Popper himself.[3] But along with this customary interpretation of Popper's work there is a second strain which constitutes a much sharper (although by no means complete) break with logical empiricism and has much in common with the new approach that we will discuss. It is because of this dual position that I refer to Popper as a transitional figure. I will begin by developing the more customary interpretation of Popper's work.

Strict Falsificationism

For Popper the central problem of the philosophy of science is what he calls the problem of *demarcation*, i.e., the problem of finding a criterion by which we can distinguish scientific theories from metaphysics and pseudo-science.[4] At first glance this might appear to be essentially the same starting point as that of the positivists, but for Popper a demarcation criterion is not a theory of meaning and metaphysics is not meaningless. Popper does not consider the problem of meaning to be a serious problem and in seeking a demarcation criterion he is attempting only to delimit one area of meaningful discourse: science.[5]

The criterion of demarcation that Popper finds implicit in the work of the positivists is one which might be called "verificationism:" the distinguishing characteristic of scientific propositions is that they can be confirmed by experience. This, as we have seen, must be subdivided into two views: the earlier thesis, held for example by Wittgenstein and Schlick, that complete verification of scientific claims is possible, and the later view of writers such as Carnap, Hempel and Reichenbach that experience can confirm scientific propositions in the sense of showing them to be probable. Popper rejects both these forms of verificationism, and indeed any attempt to construct an inductive logic. His main objections to inductive logic are the traditional ones. On the one hand, inductive inferences are not logical inferences in the only sense of "logic" that Popper will admit, i.e., tautological transformations such as we find in the deductive logic of *Principia Mathematica*. The crucial characteristic of such transformations is that the conclusion of an argument can have no greater content than the premises, but no attempt to demonstrate a universal proposition on the basis of premises which consist of a finite set of singular propositions can ever be a logically valid argument in this sense. On the other hand, if we interpret inductive arguments as making use of some synthetic principle of induction, then this principle itself must be justified and, unless we accept some form of a priori justification of induction, which no empiricist is prepared to do, we must attempt to justify the principle of induction inductively. But then the argument either becomes circular or leads to an infinite regress of principles of induction. Attempts to modify the inductivist thesis and hold that induction only shows the conclusion to be probable fall, according to Popper, under the same objections. This approach requires a new, appropriately modified, principle of induction which must itself be justified (i.e., shown to be probable to some degree since this principle too is neither analytic nor synthetic a priori), and so on.[6]

Having rejected the thesis that scientific propositions can be either verified or given probability values, Popper attempts to reconstruct the logic of science in such a way that deductive logic alone is sufficient for the evaluation of scientific claims. This reconstruction yields a new demarcation criterion. For while a universal propositions cannot be deduced from any set of observation statements, other propositions can be deduced from universal propositions and, in particular, observation statements can be deduced from universal propositions supplemented by appropriate statements of initial conditions and boundary conditions.[7] If one of these deduced observation statements is shown by experience to be false, it followed deductively by *modus tollens* that the universal proposition in question is false. It is because of this logical asymmetry between verification and falsification that Popper proposes his new demarcation criterion: a proposition is scientific only if it can be *falsified* by experience.[8] Let us develop this proposal by means of an example.

Consider the classic experiment in which measurements were made of the apparent locations of stars which appeared close to the solar disc during an eclipse. The experiment was undertaken in 1919 in order to test a consequence of Einstein's general theory of relativity which differed from that which followed from the Newtonian theory of gravitation. Einstein's theory (in conjunction with an appropriate theory of light) predicted that the gravitational field of the sun would bend light rays passing nearby while Newtonian theory had predicted no such deflection (or a much smaller deflection, depending on the theory of light used in the computations). By observing the apparent position of stars close to the sun and comparing this with the known position as computed from other observations made at times when their light did not pass close by the sun, the effect of the sun's gravity on light could be determined. The results of the observations were in agreement with the predictions of general relativity and contrary to the predictions of Newtonian theory. For Popper this observation constitutes a refutation of the Newtonian theory; but it does not constitute a verification or proof of general relativity, nor does it confer a probability value on general relativity (although we shall see in a moment that it does do something else which Popper calls "corroboration"). Having been shown to be false, Newtonian theory must now be abandoned, but it still remains a *scientific* theory.

It is this logical feature of being deductively falsifiable that distinguishes scientific theories. Pseudo-scientific theories such as astrology often make correct predicitons, but they are so formulated as to be capable of evading any falsification and are thus not scientific. And scientific theories must not only be empirically falsifiable, but a scientific claim must be rejected as soon as it has encountered a single falsifying instance. Thus, in *The Logic of Scientific Discovery* Popper writes:

In general we regard an inter-subjectively testable falsification as final (provided it is well tested): this is the way in which the asymmetry between verification and falsification of theories makes itself felt. . . . A corroborative appraisal made at a later date—that is, an appraisal made after new basic statements have been added to those already accepted—can replace a positive degree of corroboration by a negative one, but not *vice versa*.[9]

More recently, in a discussion of the light deflection observations, Popper writes, "If observation shows that the predicted effect is definitely absent, then the theory is simply refuted."[10] Similarly, a bit later in the same essay Popper writes concerning psychoanalytic theory, "*criteria of refutation* have to be laid down before hand: it must be agreed which observable situations, if actually observed, mean that theory is refuted."[11] If the proponents of a theory attempt to protect it from falsification by stratagems such as the addition of ad hoc hypotheses or reinterpretation of theoretical postulates as definitions (moves which are always logically possible), they thereby make the theory unfalsifiable and thus, on Popper's demarcation criterion, remove its status as a *scientific* theory.

There is another way of putting the falsifiability thesis which is also illuminating. In the notation of principia logic, any universal proposition such as "$(x) (Px \supset Qx)$" is logically equivalent to the negation of an existential proposition: "$\sim (\exists x) (Px \cdot \sim Qx)$." What the latter proposition says is that a certain type of empirical situation, a situation in which a single object is both P and not Q, cannot occur. The discovery of a single object which is both P and not Q provides us with a premise, "$Pa \cdot \sim Qa$," from which we can deduce the falsity of the universal proposition no matter how many instances of objects which are both P and Q we have already observed. From this point of view, universal propositions are best construed as statements of prohibitions, as forbidding the occurrence of certain empirical situations,[12] and the range of situations that a theory forbids can be taken as a measure of its empirical content: the more a theory forbids, the more it says and the more it says, the greater the chances of its being refuted.[13] This analysis throws further light on Popper's objections to the probability version of inductive logic, since the important scientific theories are those which have the greatest empirical content and are thus the least probable.[14] Science does not progress as a result of scientists attempting to play it safe by offering hypotheses which stick as close as possible to the available evidence. Rather, science progresses as a result of scientists making bold conjectures which go beyond the available data; having made his conjectures, the scientist's prime concern in testing his theories is not to attempt to prove them true, but to attempt to refute them.[15]

We have seen that according to Popper the discovery of instances which are in accordance with the predictions of a theory neither confirm the theory nor confer a degree of probability on it, but it does not follow from this that they are totally irrelevant to the evaluation of the theory; under certain circumstances they serve as corroborating instances. A theory is corroborated whenever it passes a test, i.e., whenever an observation whose outcome could possibly have refuted the theory fails to refute it.[16] How much a particular test tends to increase the degree of corroboration depends on the severity of the test. Passing a severe test, one in which an outcome favorable to the theory is highly unlikely, increases the degree of corroboration more than does passing an easy test, but in contrast to those advocates of inductive logic who hold that we can assign numerical probability values to scientific hypotheses, Popper maintains that "we cannot define a numerically calculable degree of corroboration, but can speak only roughly in terms of positive degrees of corroboration, negative degrees of corroboration, and so forth."[17]

According to this interpretation of Popper, then, the history of science consists of a series of conjectures and refutations. The scientist's job is to offer conjectures, hypotheses which have no logical foundation at all, and then attempt to refute them. The process of refutation consists of deducing observable results from our theory (in conjunction with appropriate initial and boundary conditions) and then deducing the falsity of our conjectures when the predicted observable results are shown not to be the case. The only logic of science is deductive logic; all other factors which may enter into scientific research are alogical and thus irrelevant to the philosopher of science whose concern is the "logic of knowledge,"[18] although they may form a part of the subject matter of empirical sciences such as psychology and sociology.

The degree to which this analysis of science is squarely within the presuppositional framework of logical empiricism should be clear. To begin with, Popper takes the problems of the philosopher of science to be logical problems and the tautological transformations of *Principia Mathematica* to be the canon of logic. While he rejects the attempt to construct a theory of confirmation, he does so for a perfectly respectable logical empiricist reason: that no adequate inductive *logic* can be constructed. Similarly, Popper is at one with the logical empiricists in holding that the objectivity of science derives from its being constructed on an "empirical basis." The empirical basis consists of singular existential propositions which Popper calls "basic statements,"[19] the familiar propositions of the form "Px," which tell us that a particular thing or event is in a particular region of space-time. These propositions are accepted as a result of observation, and they serve as premises for the refutation of proposed theories and as the grounds for accepting a theory as corroborated when attempts at refutation fail. But let us now look at the other side of the Popperian coin.

Basic Statements   The second interpretation of Popper, which has come into prominence in recent years,[20] can best be approached through a reexamination of the epistemic status of basic statements and of the role that they play in the falsification process. We will begin with the former question, continuing to assume, for the moment, that basic statements serve as the premises in falsifying arguments.

Clearly, the strength of any particular falsification depends on the epistemic status of the basic statements, since Popper's entire approach is built on the fact that there is a logical relation which permits us to infer the negation of a universal statement from a singular premise. In view of Popper's continual emphasis on the conclusiveness of the *modus tollens* argument, it is not surprising that he has so often been interpreted as holding that falsifications are in all cases final. But the conclusiveness of the *modus tollens* argument is not, by itself, sufficient to establish the finality of any falsification. In order to accomplish this, the basic statements which serve as the premises of falsification arguments must themselves be established with finality. Popper denies that this can be done; indeed, to hold that basic statements could be known indubitably would be inconsistent with his entire methodology. Let us consider the main reasons why Popper does not and cannot admit any finally established basic statements into his philosophy of science.

To begin with, Popper points out that no strict disproof of a scientific theory is possible because experimental results can always be challenged.

> In point of fact, no conclusive disproof of a theory can ever be produced; for it is always possible to say that the experimental results are not reliable, or that the discrepancies which are asserted to exist between the experimental results and the theory are only apparent and that they will disappear with the advance of our understanding. (In the struggle against Einstein, both these arguments were often used in support of Newtonian mechanics, and similar arguments abound in the field of the social sciences.) If you insist on strict proof (or strict disproof) in the empirical sciences, you will never benefit from experience, and never learn from it how wrong you are.[21]

But if it is always possible to question experimental results, then no basic statement can be established with finality. More importantly, if we can always evade falsification on the grounds that an established counter-instance will be shown by further research to be only an apparent counter-instance (as has happened in a number of important cases[22]), there is no final refutation of a theory.

There is, for Popper, a second reason, considerably stronger than the fact that experimental results can always be questioned, which prevents the conclusive establishment of any basic statements. Basic statements are accepted or rejected as a result of experience, but it is logically impossible for experience to prove or disprove any statement. Recall that the only notion of "proof" that Popper admits is that of logical deduction, and logical relations only hold between statements. But experiences are not statements, they are psychological events and no *logical* relation can hold between a statement and a psychological event. There is still a close tie between experience and basic statements; indeed, Popper maintains the empiricist position that experience must supply the basis for all scientific theories and that it is the basic statements that provide the empirical basis of the testing process. But experiences can only motivate our acceptance of basic statements, they cannot prove these statements.[23]

The third, final reason why basic statements cannot be conclusively established is perhaps the most important, since it is central to Popper's entire approach. Since basic statements enter into scientific arguments, they must be scientific statements, so that, in accordance with Popper's demarcation criterion, they must be falsifiable. To hold that science rests on some set of indubitable observation reports, as many empiricists have held, is, for Popper, to make science rest on a non-scientific foundation. The view of science as a set of conjectures and refutations applies to all strata of science from the lowest level report of experimental results to the most complex theory. *All* scientific statements are falsifiable conjectures.

How, then, can falsification be accomplished? For Popper falsification takes place only after scientists agree to accept a basic statement as adequately corroborated.

> Every test of a theory, whether resulting in its corroboration or falsification, must stop at some basic statement or other which we *decide to accept*. If we do not come to any decision, and do not accept some basic statement or other, then the test will have led nowhere. But considered from a logical point of view, the situation is never such that it compels us to stop at this particular basic statement rather than at that, or else give up the test altogether. For any basic statement can again in its turn be subjected to tests, using as a touchstone any of the basic statements which can be deduced from it with the help of some theory, either the one under test, or another.[24]

Again, "From a logical point of view, the testing of a theory depends upon basic statements whose acceptance or rejection, in its turn, depends upon our *decisions*. Thus it is *decisions* which settle the fate of theories."[25] Put somewhat differently, since the acceptance of a basic statement rests on a decision by the scientists concerned rather than on some form of proof, an accepted basic statement is a convention. But Popper maintains that his philosophy is distinguished from the conventionalism of writers such as Duhem and Poincaré in that for the latter the acceptance of universal propositions is determined by convention while for him it is the acceptance of singular propositions that is determined by convention.[26]

Given this analysis of basic statements, it would seem that a fundamental reconsideration of the thrust of Popper's falsificationism is in order. In particular we must ask whether there are any important differences between the process by which a theory is falsified and the process by which it is corroborated. It does, of course, remain true that once we have accepted an appropriate set of basic statements we can formally refute a theory, while we can never formally prove a theory. But while this may be an important difference from a strictly logical point of view, it looses much of its significance once we recognize the tentative nature of basic statements. The scientist may, within the framework of Popper's methodology, always legitimately choose to try to refute unwanted basic statements rather than use them to refute theories. And although basic statements do have a privileged status for Popper in that they are conventions accepted as a result of experience, in any case in which a basic statement is questioned it would seem that the process of refutation of a theory must be suspended until the basic statement involved has been tested and *corroborated*. Thus in at least some cases the falsification of one scientific conjecture requires the prior corroboration of another scientific conjecture.

At this point the status of basic statements has become thoroughly ambiguous, and because of their fundamental role in Popper's methodology that methodology itself has become ambiguous. Either basic statements are unfalsifiable conventions and the "empirical basis" of science is itself nonscientific, or they are falsifiable conjectures and the scientist has as much of a duty to attempt to refute them as he has to attempt to refute any other scientific conjecture. But if we adopt the latter option the conclusion we just reached can be restated in an even stronger form, since it follows that in any case in which a basic statement is used as a premise of a refutation, the duty to attempt to refute the basic statement is equivalent to a duty to attempt to defend the theory. Once this is recognized, the persistent Popperian rhetoric about the scientist's duty to attempt to refute his conjectures loses much of its force.

This conclusion receives further support if we turn to the second issue that I proposed to discuss: the precise role that basic statements play in the falsification process. For while Popper maintains that the falsification of a theory requires the acceptance of a basic statement which contradicts it, he also maintains that "this condition is necessary, but not sufficient"[27] for falsification. Popper is concerned to eliminate any suggestion that a scientific theory may be falsified as a result of "a few stray basic statements"[28] contradicting it, basic statements which may be the result of errors or accidents. Rather, a theory is falsified only after we have established

> a *reproducible effect* which refutes the theory. In other words, we only accept the falsification if a low-level empirical hypothesis which describes such an effect is proposed and corroborated. This kind of hypothesis may be called a *falsifying hypothesis*. . . . The rider that the hypothesis should be corroborated refers to tests which it ought to have passed—tests which confront it with accepted basic statements.[29]

And again:

> It has already been briefly indicated what role the basic
> statements play within the epistemological theory I advocate.
> We need them in order to decide whether a theory is to be
> called falsifiable, i.e., empirical. And we also need them for
> the corroboration of falsifying hypotheses, and thus for the
> falsification of theories.[30]

Thus, for Popper the premises of falsifying arguments are not basic statements, but falsifying hypotheses that have been corroborated as a result of tests against basic statements. It thus turns out that no actual falsification can occur until after a falsifying hypothesis has been corroborated so that in spite of the logical asymmetry between verification and falsification, *every* case of falsification requires a prior corroboration and in no particular case can a falsification be any stronger or more final than a corroboration.

At this point it is by no means clear exactly what the methodology of Popper's *Logic of Scientific Discovery* is, although it is quite clear that is has moved a long way from the approach of the logical empiricists. In particular, it is unclear how the scientist is to decide in any particular case in which a low level hypothesis contradicts a theory whether to reject the theory or to attempt to defend it by seeking a refutation of the hypothesis. The approach we will discuss in Part II will have no difficulty with this kind of situation, since the structure of such decisions will be analyzed in terms of the informed judgment of the individual scientist and the relevant scientific community, but this is a long way from Popper's demand for clear methodological rules and his attempt to construct a purely deductive logic of science. On the other hand, the ambiguities in Popper's work can at least help us to explain how a philosopher such as Lakatos can call himself a Popperian while developing a philosophy of science in which the role of experiment and observation is reduced to an absolute minimum[31] and a philosophy of historical method which makes it acceptable to ignore contrary evidence.[32]

In the light of the above discussion it is well worth asking why Popper and many of his disciples continually put such great emphasis on the role of *modus tollens,* the privileged status of falsification over corroboration, and the special role of basic statements in the methodology of science. Once again I suggest that the reasons for this emphasis are to be found in the lingering remains of the logical empiricist presuppositions. In particular in the logical presupposition of the special role of formal logic in the philosophy of science, and secondly in the empiricist presupposition that the objectivity of science is completely derived from its appeal to observation or, at least, to some special set of statements which have a particularly close tie to observation. Once we free ourselves from these two presuppositions a very different picture of the nature of science emerges, one in which the judgment of the scientific community plays a much greater role than does the application of formal rules and effective criteria, and in which theory and observation are much more nearly co-equal partners in the construction of science. Popper played an important role in moving the philosophy of science in this new direction, but he did not complete the transition himself.

There is one more point that is worth making before we leave our discussion of Popper since it provides a striking illustration of the pervasive role of presuppositions in human thought. Much of this chapter was devoted to contrasting what might be called the two different Popperian philosophies: strict falsificationism which was the dominant interpretation of Popper until relatively recently and to which, as we have seen, Popper himself has at times subscribed, and a much weaker modified falsificationism that follows from the full analysis of basic statements. It is only recently that this latter Popperian philosophy has come into the central focus of methodological discussions in spite of its solid foundation in the text of his *Logic of Scientific Discovery* and if, as Popper maintains, the view that he once was a strict falsificationist is a misinterpretation,[33] it is by no means the most serious misinterpretation to which his work has been subjected. Many writers have taken his demarcation criterion to be a *criterion of meaning* which he offered as an alternative to the positivists verification theory of meaning.[34] What is interesting here is that these "misreadings" of Popper were made at a time when the philosophy of science was completely dominated by logical empiricism, and they were misreadings which tended to bring him into closer agreement with the spirit of logical empiricism. During the past fifteen years or so a new approach to the philosophy of science emerged, many of these "misinterpretations" were corrected and Popper has been reread in a way which brings him considerably closer to the new approach. Of course the existence of two different interpretations of his work is not due solely to its having been read from the point of view of different philosophies of science; the possibility of different readings arises from ambiguities in his writings. But given these ambiguities, the prevailing intellectual climate has played a dominant role in determining what philosophy of science was associated with Popper's name and texts.

**Conclusion: Toward a
New Understanding**

Our main concern in part I has been to formulate the presuppositions of logical empiricism and to show how these presuppositions control logical empiricist philosophy of science, and in particular how they generate the problems which the philosopher works on and his criteria for what constitute acceptable solutions. We have not examined all the problems of logical empiricism, but we have dealt with the central issues and we have considered enough of the problems and the literature to demonstrate our thesis.

The full thrust of this thesis can best be appreciated by reflecting on the relation between the logical empiricists' view of science and their view of the philosophy of science. One of the central empiricist claims about science is that it rests on what Feigl calls the " 'soil' of observation"[35] and has no presuppositions. Since the logical empiricists consider themselves to be scientific philosophers, it follows that they view their own philosophy of science as presuppositionless. Just as, in their view, science begins from the solid data of observation, so their own analyses of science begin from the solid body of principia logic and scientific empiricism. We have found adequate reasons for doubting that logical empiricism is a presupposition-free philosophy; we will, in Part II, find equally strong reasons for doubting that science itself is free of presuppositions, but the order of our argument in Part II must be the reverse of that in Part I. Logical empiricist philosophy of science is a philosophic research program which grew up in the context of an already well developed epistemology, but the new approach to the philosophy of science that we now turn to has a different history. It emerged largely in response to the growing sterility of logical empiricism, its failure to achieve adequate solutions to its own problems and to further clarify the nature of science, as well as from the many anomalies raised by new work in the history of science. There has been no clear formulation of the epistemological framework of the new approach and it is highly doubtful that there is any one epistemology to which all its advocates would subscribe. Indeed, some of these writers continue to consider themselves empiricists,[36] while some (not necessarily the same ones) accuse others of being idealists, intending this to be a devastating criticism.[37] As a result, in Part II we will first develop at length the new image of science. Only after we have done this, in the final chapter, will we attempt to formulate the epistemology implicit in the new work.

Part II   **The New Image of Science**

Chapter Six

Perception and Theory

One of the cornerstones of logical empiricism is the thesis that there is a fundamental distinction between uninterpreted scientific theories and the body of perceptual experience which confers meaning on our theories and determines which ones are to be accepted. The system of postulates which make up a theory " 'floats' or 'hovers' freely above the plane of empirical facts,"[1] but it is the empirical facts, which are known independently of any theory, that guarantee the objectivity of science. One starting point of the new philosophy of science, however, is an attack on the empiricist theory of perception. In response to the view that perception provides us with pure facts, it is argued that the knowledge, beliefs and theories we already hold play a fundamental role in determining what we perceive. In order to understand the thrust of this approach and to clarify some of the new problems it generates, we will have to reexamine the role that perception plays in our knowledge. For the most part we will confine our discussion to cases of visual perception since this is the most important form of perception in scientific research.

Consider a relatively common, everyday instance of perception such as seeing my typewriter. Now in order to see that this object is a typewriter it is not sufficient that I just look at it; it is necessary that I already know what a typewriter is. Simply glancing at objects with normal eyesight will undoubtedly stimulate my retina, initiate complex electro-chemical processes in my brain and nervous system, and perhaps even result in some conscious experience, but it will not supply me with meaningful information about the world around me. In order to derive information from perception it is necessary that I be able to identify the objects that I encounter, and in order to identify them it is necessary that I already have available a relevant body of information. As Hanson puts it, "there is more to seeing than meets the eye."[2] We will refer to cases in which we do gain information via perception as "significant perception," and this is the only form of perception that will concern us.

The point holds for all the common objects we encounter as well as their properties, although we often fail to note the role that our knowledge and beliefs play in the recognition of objects because of the great familiarity of the objects of our everyday experience and because so much of the information necessary for us to recognize them is learned through the largely unreflective process of growing up in a culture. But a bit of reflection should be sufficient to show how large and subtle the fund of information is that allows us, without noticing that we are doing anything special, to distinguish delivery trucks from fire trucks and paper towels from napkins.

The thesis holds even more clearly for cases of scientific perception. The ability to recognize such objects as a cathode ray tube, a red corpuscle, or a carbonaceous chondrite requires a great deal of highly specialized knowledge, and in the process of gaining that knowledge we are also learning to see the objects. Consider the following classic example taken from Duhem:

> Go into this laboratory; draw near this table crowded with so much apparatus: an electric battery, copper wire wrapped in silk, vessels filled with mercury, coils, a small iron bar carrying a mirror. An observer plunges the metallic stem of a rod, mounted with rubber, into small holes; the iron oscillates and, by means of the mirror tied to it, sends a beam of light over to a celluloid ruler, and the observer follows the movement of the light beam on it. There, no doubt, you have an experiment; by means of the vibration of this spot of light, the physicist minutely observes the oscillations of the piece of iron. Ask him now what he is doing. Is he going to answer: "I am studying the oscillations of the piece of iron carrying this mirror"? No, he will tell you that he is measuring the electrical resistance of a coil. If you are astonished and ask him what meaning these words have, and what relation they have to the phenomena he has perceived and which you have at the same time perceived, he will reply that your question would require some very long explanations and he will recommend that you take a course in electricity.[3]

In order to see what is being done in the laboratory I must understand the relevant body of physical theory; if I do not have this knowledge I cannot see that the scientist is measuring electrical resistance, nor see an electric battery, nor see a ruler, no matter how healthy my eyes are.

This conclusion will undoubtedly strike many readers as being somewhat strange. Surely, it will be suggested, the physicist is measuring electrical resistance and when I look at him this is what I see whether I recognize it or not. But while there is a sense in which this objection is well taken, it is more important for the moment to note that irrespective of what the scientist is "really" doing, what we *learn* by observing him is not determined solely by what he is doing but also depends on what the observer already knows. An observer who lacks the relevant knowledge will not gain the same information by watching the experiment as will a trained physicist and there is thus an important respect in which the layman and the physicist see different things when they observe the same experiment. Similarly, a chemist in the neighborhood of a steel mill will smell sulphur dioxide and gain a great deal more information about what is happening to his body and his environment than will a child who only smells rotten eggs. Both observers smell the same thing, a point we will examine more closely later, but what I wish to emphasize at the moment is the differences in the information the two perceivers gain from a single perceptual situation; these differences are of fundamental importance for understanding the nature of significant perception and thus for understanding the way in which perception can contribute to knowledge.

If our knowledge and beliefs play a central role in determining what we perceive, then the scientific theories that a scientist holds should play the same sort of role in determining what he observes in the course of his research; to borrow another phrase from Hanson, scientific observation should be "theory-laden." But if this is the case, it should be possible for two scientists who hold different theories to look at a single object and perceive different things. Let us examine one of Hanson's examples.

> Consider two microbiologists. They look at a prepared slide; when asked what they see, they may give different answers. One sees in the cell before him a cluster of foreign matter: it is an artefact, a coagulum resulting from inadequate staining techniques. . . . The other biologist identifies the clot as a cell organ, a 'Golgi body'. As for technique, he argues: 'The standard way of detecting a cell organ is by fixing and staining. Why single out this one technique as producing artefacts, while others disclose genuine organs?[4]

What do the two scientists disagree about? Do they disagree about what they see or only about the proper description of something which both see? If we accept the latter alternative, it follows that the undescribed thing which both scientists see plays no role in scientific knowledge nor in the resolution of scientific debates. It is the described or theory-laden percept that the two biologists are arguing about; even if we were to concede that there is some pure, theory-free datum that both perceive, no further observation of this datum would be relevant to the resolution of their disagreement.[5] Another example will help to clarify and develop this point.

Consider Kepler and Tycho Brahe looking at the sun.[6] Kepler tells us that the sun is a stationary body around which the earth moves, Brahe that it is a body which moves around the stationary earth. Now is it correct to say that they see different things, Kepler a moving body and Brahe a stationary body, or shall we say, rather, that they see the same thing, the sun, but describe it differently or make different claims about it? A philosopher who accepts the latter position is faced with the task of clarifying just what it is that they both see. Simply to say that both see the sun tells us nothing since Kepler and Brahe disagree about what the sun is. One might be tempted to reply here that, irrespective of what anyone believes, the sun is stationary and the earth moves around the sun so that when we say that they both see the sun, we are saying that they see a stationary object and that not only do they see the same thing, but Kepler describes it correctly and Brahe describes it incorrectly.[7] It is important to remember, however, that the confidence with which a contemporary can make this claim derives from hindsight, from our knowledge that it is Kepler who won this dispute. But neither Kepler nor Brahe had our perspective on this question and if we are to attempt to understand the nature of scientific disputes and the ways in which they are resolved, we must examine them from the point of view of the disputants. When we are dealing with a contemporary dispute we do not have the advantage of hindsight that we have in the case of Kepler and Brahe, and if the analysis of cases from the history of science is to throw any light on the nature of contemporary disputes, we must approach historical disputes as if they were contemporary and not allow more recent information to intrude into our analysis.[8] In terms of the information they had available, then, when Kepler saw the sun, he saw the stationary center of the universe around which the earth revolved. Brahe, on the other hand, saw a celestial body which moved around the stationary earth.

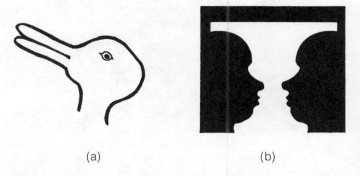

(a)   (b)

Figure 2   (a) Duck/rabbit;   (b) Faces/vase

Another kind of example which has been widely discussed and will help to throw light on the thesis that a single thing can be seen in different ways is the gestalt shift, i.e., figures such as the duck/rabbit or the faces/vase which can appear to a single observer in either of two (or more) ways. These figures were originally used by gestalt psychologists in their attack on the constancy hypothesis, i.e., the claim that what we see is entirely determined by the retinal image. The shifting figures provide a clear case of a situation in which an observer sees two different objects while the pattern of retinal stimulation remains the same. For our present purposes a discussion of these figures has the advantage of allowing a single observer to come into direct contact with a kind of situation which, in our other examples, occurs only between two different observers. In the duck/rabbit case, for example, I see what are clearly two different things, a picture of a duck and a picture of a rabbit, while being conscious of looking at the same thing throughout. At the same time, there is no basis at all for maintaining that one of the available descriptions, the duck or the rabbit, is *really* the correct one.

One might be tempted to respond that, although we do alternately see a duck and a rabbit, these are different ways of seeing a single figure and by properly directing our attention we can discover the object that is really there. Kuhn, for example, succumbs to the temptation to present this sort of analysis. Discussing the duck/rabbit case, he writes:

> The subject of a gestalt demonstration knows that his perception has shifted because he can make it shift back and forth repeatedly while he holds the same book or piece of paper in his hands. Aware that nothing in his environment has changed, he directs his attention increasingly not to the figure (duck or rabbit) but to the lines on the paper he is looking at. Ultimately he may even learn to see those lines without seeing either of the figures, and he may then say (what he could not have legitimately said earlier) that it is these lines that he really sees but that he sees them alternately *as* a duck and *as* a rabbit.[9]

Kuhn's argument turns on a distinction between two forms of seeing: simple "seeing" and "seeing as." When I see an object *as* something, it is a case of significant perception: the object is identified and thus the preception is theory-laden or, perhaps more accurately in the present case, concept-laden perception. Kuhn's suggestion seems to be, then, that if we can succeed in stripping off the identification, we make a transition from theory-laden to non-theory-laden perception, from "seeing as" to "seeing," and thus succeed in observing the object itself. But if neither the picture of the duck nor the picture of the rabbit can be what is really there because they are theory-laden percepts, Kuhn's own argument provides no reason for taking the perceived lines to be any less theory-laden. As he points out, we must learn to see the lines, and objects that we have learned to see stand as paradigm cases of theory-laden percepts. Similarly, by properly directing our attention we can also learn to see an area on the paper and this would provide us with yet a fourth way of seeing the figure, but there certainly seems to be no reason why we should take "lines" to be a better description of what we really see than "an area."

Still another question must be raised here: Why should we begin to look for the lines in the first place? Kuhn's analysis could be read as suggesting, not that we discover what is there by stripping off layers of concepts, but that we first know from some other source that lines are what is really on the paper and then attempt to see them. Looked at in this way, concept-free seeing is not the criterion for what is really there, but rather our independent knowledge of what is really there tells us what to look for. But what is the source of this knowledge? It would seem that it must lie in some theory that we hold or information we have about the diagram. But then our seeing the lines is clearly theory-laden seeing and we might even go so far as to suggest that, given the relevant theories, we come to see the lines *as* what is really there. Given a different set of theories or a different body of information (for example, information about who drew the diagram and what his intentions were), we might then see the rabbit as what is really there. In any case, all of the ways in which we can see this object and relate it to what we know require that the object be viewed in terms of our knowledge and are thus cases of "seeing as."

The use of the "seeing as" locution serves to indicate that we are dealing with the perception of identified objects and thus with significant perception. Cases in which we recognize not only objects but facts about objects or situations are often denoted by the use of "seeing that" expressions. An analysis of cases of "seeing that" will carry our examination of significant perception considerably further.[10]

Consider once again the case of the physicist measuring electrical resistance. If we draw a distinction between "seeing" and "seeing that" we have the linguistic machinery to say that while anyone who observes the experimenter will indeed *see* a physicist measuring electrical resistance, only one who already has the necessary knowledge of physics will *see that* he is measuring resistance. For an observer ignorant of physics the observation adds nothing to his knowledge. Similarly, I can look at my watch and *see that* it is noon while a child who has not yet learned to tell time cannot *see that* it is noon at all, although he might well be able to *see that* I am wearing a wrist watch whereas a still younger child will not even recognize this much. Perhaps the most revealing cases of "seeing that" are those in which I notice the absence of an object. If someone has, for example, removed my typewriter from my study, I will, upon entering my study, immediately see that it is missing, while another person who has never been in my study will not see that the typewriter is missing no matter how healthy his vision is.

In every case in which I see that something is the case I gain information as a result of my seeing, but what information I gain depends not only on what visual events take place in my eyes, nerves and brain, but also on what information I bring with me. And the converse holds as well. Every instance in which I gain information as a result of seeing is an instance in which I *see that* something is the case and the more I already know about the situation in question the more I can learn.[11] There are, as Hanson notes, cases in which we cannot immediately bring our knowledge to bear on the objects that we see.

In microscopy one often reports sensations in a phenomenal, lusterless way: 'it is green in this light; darkened areas mark the broad end. . . .' So too the physicist may say: 'the needle oscillates, and there is a faint streak near the neon parabola. Scintillations appear near the periphery of the cathodescope. . . .' To deny that these are genuine cases of seeing, even observing, would be unsound, just as is the suggestion that they are the *only* genuine cases of seeing.[12]

But as long as these observations remain on a purely phenomenal level, they do not become a part of our knowledge and the very possibility of their becoming relevant to our knowledge depends on their already being related to some body of information. For the observer in these cases knows that he is looking through a microscope or using a cathodescope and what their uses are, and undoubtedly has a great deal of other information about the objects he is studying and the tests he is carrying out. It is this information that allows him to recognize that an unfamiliar event has occurred and it is in terms of this information that he attempts to identify the streaks and patches he observes and eventually integrate them into what he knows.

On the other hand, we must not go too far and forget that despite the crucially important role of our knowledge in determining what we observe, observation only occurs in those cases in which we make use of our physical senses. Hanson, at times, comes very close to forgetting this. At one point he writes:

A blind man cannot see how a timepiece is designed, or what distinguishes it from other clocks. Still, he may see that, if it is a clock at all, it will embody certain dynamical principles; and may explain the action to his young apprentice. The latter, however keen his vision, can only describe the perturbations of the clock. . . .[13]

The sense in which we say that the blind man can see that the clock works in a particular way is not the sense of "seeing that" that concerns us here. Our interest is with the keen eyed apprentice who can see that the clock works in a particular way only after the principles have been explained to him.

Now let us carry our analysis of the objects of significant perception one step further. Our discussion has focused on two kinds of situations: cases in which different perceivers with different information learn different things from observing a single object, and cases in which one of the observers is completely lacking the relevant knowledge and so learns nothing at all from his observation. We can describe both of these situations by saying that in the first case the observed objects have a different meaning for the different observers and in the second case the objects in question have no meaning at all for the uninformed observer. It is, then, the meaning of the observed situation that becomes a part of our knowledge and the objects of significant perception are thus *meanings*.[14] Hanson comes very close to drawing this conclusion, although he does not take the final step:

There is no more in the east at dawn for the scientist to see than there is for the lunatic to see. And yet the scientist sees incomparably more. The objects of our seeing, hearing, touching, tasting and smelling acquire meaning for us, only when we can link up what is directly given in experience with what is not. A brilliant white spot of light against a deep blue background has an ineffable, incommunicable, and very personal quality. After all, the impression itself is not different in type from what I see after colliding with a football or with someone's fist. But in normal circumstances such a spot of light can be seen as a star, can *mean* a star situated within a certain region of the heavens many light-years away.[15]

Since the suggestion that we perceive meanings will undoubtedly strike many readers as rather strange, it will be useful to compare the kinds of observation we have thus far considered with the case of reading a text, since the latter is clearly a case of visual perception which involves meanings on some level. When I read a text I see, but the object of my attention and the information I gain from my reading is the meaning of the text. This meaning depends on a number of factors. To begin with, there is the text itself: I read only if I actually see the text and if the meaning that I find there does have some foundation in it. But being able to see the text is not a sufficient condition for reading. I must also be able to read the language in which it is written and, often, I must have some knowledge of its subject. A normal adult who knows nothing about, say, geology cannot read an advanced geology text even if he recognizes almost every word; the text will have no meaning for him. (He may have particular difficulties with some of the technical terms, but consider how much geology he would have to learn in order to learn the meanings of these terms. A dictionary alone will not suffice.) But even if there are cases in which no knowledge of a special subject is required, the meaning that I find in the text is still dependent on my knowledge of the language and of the context. The meaning of "lead" or "wound," for example, depends on the context; the meaning of "man" is different when it is printed in an English text and in a German text, and all of these signs are meaningless in French. Again, any collection of letters is a possibly meaningful word in some possible language and is capable of bearing any amount of meaning in a code. In general, then, to read a text is to become aware of its meaning. I cannot become aware of the meaning of a text if I do not have the relevant information, and a given text can have different meanings for different readers and no meaning for some: a situation consistent with what we have found in other cases of visual perception.

It will no doubt be replied here that the parallel with reading does nothing to support the claim that the proper objects of perception are meanings, for while the object of our attention when we read is always a complex of meanings, it is still not the meanings that we *see*. There are two responses that may be made to this objection. The first is to recall that we are concerned with perception as a source of information: whatever it is that we "really see" when we are reading, it is only the meaning of what we see that can become part of our knowledge, just as when I observe familiar objects or laboratory phenomena it is only the meaning of these objects that is relevant to what we know. If there are bare, meaningless data, the very fact that they are meaningless makes them non-significant and irrelevant to knowledge.

But a much stronger reply is also available. We can take up the claim that we do not see meanings when we read and attempt to clarify just what it is that we do see. Clearly responses such as that we see words or letters or even ink marks on paper will not do, since in order to recognize any of these objects I must see *that* I have a word or a letter or an ink mark in front of me, I must recognize the object, and I am again aware of a meaning. Our opponent may retreat to the distinction between "seeing" and "seeing that" and maintain that whether or not I recognize that the objects before me are, say, ink marks, they still are ink marks and thus it is ink marks that I see. But this raises the further question of how one is to discover that these are indeed ink marks, since this line of argument requires that I know what is really on the paper before I can determine what it is that I see. Now there are many ways in which I can discover this: by chemical analysis, by asking the publisher in the case of a printed text or the author in the case of a written manuscript, or even by simply looking at the text and observing that it is printed in ink (assuming, of course, that I am sufficiently familiar with printed texts to make this observation). But whatever method I adopt, it is clear that the observation of unrecognized, meaningless visual data is not among them.

Indeed, we can see the full weakness of the datum thesis by carrying to its logical conclusion the retreat from the observation of meanings to the pure observable. In order to draw a distinction between the meaning we become aware of and the pure observable which it is claimed we see, the pure observable must itself be an unrecognized object; it cannot even be an object which is recognized as being unfamiliar. But those who wish to make this distinction must then explain how the observation of an unrecognized object, whether on the printed page, in our everyday lives, or in the scientist's laboratory, can serve as the basis for becoming aware of a meaning. I do not think that this can be done and I submit that in all cases of significant perception it is only meanings that we are concerned with and meanings that we see.

Thus far we have argued that meanings are the objects of significant perception. Now let us attempt to show directly that sense-data, assuming such things exist and that we are conscious of them, cannot be the primary objects of our knowledge. For the empiricist, data constitutes the rock bottom stratum of our knowledge. But if this is the case, then all sense-data must be equally important and must play equal roles in our knowledge, there being no more fundamental level of knowledge to which we can appeal in order to sort out or select our data. Clearly this is not the case. In my everyday experience and activities I continually select out as meaningful only a very small number of the sensations which appear in my visual field, my auditory field, etc. and it is only to these sensations that I pay attention. While driving my car, for example, I pay attention to other cars, traffic signals, pedestrians, and road markers that are relevant to driving, while ignoring the trees that line the road, minor blemishes on the road surface, and specks of dust on my windshield and on the lenses of my glasses. Similarly, I may hear an officer's whistle, the warning bells at a railroad crossing, the sound of a collision, and perhaps the announcer's voice on my radio while ignoring the thousands of other sounds of a busy street. In all these cases, including those in which the ignored objects are recognized or recognizeable, my knowledge of the activity I am engaged in is epistemically more fundamental than the data I observe in that the demands of this activity determine what data I pay attention to and what data are ignored as meaningless.

This holds even more decisively in the case of scientific observation. The scientist does not record everything he observes but rather only those things which the theories he accepts indicate are significant. A physicist examining cloud chamber photographs, for example, will ignore any number of streaks as spurious marks due to dirt or to scratches on the photograph. Similarly, the physicist does not include the shape of the laboratory table, or the color of the walls, or the number of cigarette butts on the floor, or his dreams of the preceding night among his observational data (although his analyst will be much more interested in the last of these than in the cloud chamber observations). Finally, we should note that in the important case in which a scientist identifies a phenomenon as anomalous or problematic he is clearly observing its meaning in terms of the theories he holds, for if he had no beliefs about what ought to occur in the situation in question, no occurrence could be perceived as problematic.[16] In all these cases it is the available body of theory that provides the criteria for determining which of the observed data are to play a role in knowledge, but our theories could not possibly play this role if our data constituted the fundamental level of knowledge on which all theories are built. Returning to the diagram which Feigl used to show the relation between the postulates of a theory and experience,[17] and assuming for the moment that this diagram does accurately show how experience and theory become related to each other, it would be more nearly correct to say that the diagram illustrates how our theories confer meaning on experience than to adopt the converse position which Feigl attempts to illustrate.

Three Problems

I have been proposing a view of the nature of significant perception that is fundamentally different from the empiricist approach, and from this point of view the problems of perception that require further research should take on a very different aspect. To illustrate this let us consider three problems. One, the problem of psychologism, is a problem for empiricist theories of knowledge; rather than offering a new solution I will argue that the problem is dissolved by the analysis of perception developed here. The other two problems I will consider, that of how two different perceived objects can still be recognized as the same object and that of relativism, are problems for our theory of perception and I will propose a solution to each of them.

Popper's formulation of the problem of psychologism is perhaps the clearest.[18] Empiricism requires that theories be tested by deducing from the theory consequences which can be compared with the results of observation. Having, on the one hand, an elaborated body of theory and, on the other hand, a relevant observation, we should be able to use the observation either to deduce the falsity of the theory or (for some empiricists) to induce some degree of confirmation for it. But an elaborated theory is a body of propositions and an observation is a psychological event; we can deduce (or induce) propositions from other propositions, but we can neither deduce nor induce propositions from psychological events nor psychological events from propositions. How, then, do observations come to play a role in scientific knowledge? Because he is concerned to avoid psychologism, Popper maintains that experience can motivate a decision to accept or reject a singular statement but can never prove or disprove any singular statement, and thus that all singular statements are accepted or rejected by convention.[19] (Logical empiricists in general tend to be less candid than Popper and to take the empirical basis for knowledge to be some form of observation statement without ever squarely facing the problem of the relation between observation statements and experience.) Clearly the entire problem derives from the presuppositions that the experience that is relevant to scientific knowledge consists of Humean impressions and that these are entities of a type fundamentally different from propositions, so that no logical relations can hold between experience and propositional knowledge (indeed, for Hume, impressions cannot enter into any *logical* relations at all). But once we recognize that the objects we are concerned with when we perceive and the objects we are concerned with when we understand a theory are both meanings and are thus of the same logical type, the putative gap between theory and observation disappears and the problem of psychologism is dissolved.

We turn now to a problem that is generated by the claim that perception is theory-laden. Central to this analysis is the thesis that a single perceiver can, at different times, see a given object in different ways and two different observers can simultaneously see the same object differently. In order for these situations to have any philosophical significance it is crucial that the observers recognize that a single object is being observed differently. If I see a duck and you see a rabbit we are undoubtedly seeing different things, but if we are looking at different pictures no puzzles arise. Now although it is quite clear that I am seeing something different when I see a rabbit than when I see a duck, how it is possible for me to be also seeing the same thing in both cases is considerably less clear. This point must be dealt with if our analysis of perception is to be complete.

Here a sense-datum approach to perception seems to be particularly appropriate for the datum theoriest can maintain that it is the associations or interpretations that accrue to my observation of the datum that account for differences in the way I see it while the sense-datum remains the same whether I am seeing a duck or a rabbit. In the case of two different observers, the datum theoriest can maintain that the observers are seeing qualitatively similar sense-data while, again, the accrued associations or interpretations differ. On the other hand, we have already seen that the datum theory is completely incapable of accounting for the ways in which observation enters into knowledge, so that its ability to resolve this one problem is of relatively little impact especially if the sameness of the percept can be accounted for equally well on the basis of our analysis of observation. Let us attempt to do this.

This phenomenon is considerably less mysterious than it has often been taken to be. The clue to an adequate analysis is given in the passage from Kuhn which was criticized earlier.[20] In the first sentence Kuhn writes that the subject sees the figure shift "while he holds the same book or piece of paper in his hands." There is an unfortunate tendency among many writers on perception to focus on one object at a time and forget that this object is always a part of a much more complex situation which is being perceived. I do not just see a free floating picture of a duck or a rabbit; rather I see the picture on a page which is in a book sitting on my desk. And while I am aware of the picture shifting from a duck to a rabbit, I am also aware that the page, the book, the desk and the room have not been altered. It is my perception of the complete situation and the information I have about situations of this kind that accounts for my awareness of sameness and my puzzlement as to why the picture shifts. If I had to turn the page in order to shift from the duck to a rabbit, or if I saw the same shift while watching a motion picture, I would not expect the sort of stability that I expect from a diagram on a page and would not experience any puzzlement at all. Put somewhat differently, when the total perceptual situation, including all I know about such situations, is taken into account, the sameness of the object before and after the shift is a part of the perceived meaning.

The same analysis can be applied to the case of the two scientists. The biologist looks through his microscope and then asks his colleague to look through the same microscope. It is because they both recognize that they are looking through the same microscope that the differences in what they perceive are puzzling. If the puzzlement is sufficiently troubling they may even double check to make sure that the same slide is still in the microscope, that the lighting has not been altered, etc., but once these tests have been passed, they have adequate grounds for believing that they are looking at the same thing. It is the total situation, then, and not some one element that can be picked out of the situation by properly directing our attention, that accounts for the awareness of sameness, and it is recognition of variations in what is already recognized as a single object that leads to sameness/difference situations.

We turn now to the problem of relativism. For the empiricist it is the appeal to theory-free observed data that guarantees the objectivity of science. It is often argued that if scientific observation is theory-laden this objectivity is destroyed, for if the theories that we already hold play a central role in determining what we perceive then it is impossible to appeal to observation to test the acceptability of scientific theories. For scientific theories to be empirically testable, the familiar argument goes, the percepts against which they are tested must be independent of the theory. A theory which creates its own data can never be refuted by this data.

We can begin our response by noting that I have nowhere maintained that theories create their own data nor that our theories *alone* determine what we perceive. Rather, the objects of perception are the results of contributions from *both* our theories *and* the action of the external world on our sense organs. Because of this dual source of our percepts, objects can be seen in many different ways, but it does not follow that a given object can be seen in any way at all. Consider again the duck/rabbit. We have already seen that this figure can be seen as a duck, a rabbit, a set of lines, or an area, and one might plausibly imagine its being seen as a piece of laboratory apparatus, a religious symbol, or some other animal by an observer with the appropriate experience. But try as I will, I cannot see this figure as my wife, the Washington monument, or a herd of swine. Unlike the Kantian position, or, rather, one interpretation of the Kantian position,[21] I do not maintain that theories impose structure on a *neutral* material. The dichotomy between the view of perception as the passive observation of objects which are whatever they appear to be and perception as the creation of perceptual objects out of nothing is by no means exhaustive. A third possibility is that we shape our percepts out of an already structured but still malleable material. This perceptual material, whatever it may be, will serve to limit the class of possible constructs without dictating a unique percept.

Again the parallel with reading is illuminating. A variety of interpretations of, for example, the *Critique of Pure Reason* have been proposed, but no matter how widely scholars differ on what is the correct reading of the text, no one can open the *Critique* and *read* the *Nichomachean Ethics* or *Moby Dick*. Clearly this approach is compatible with the kind of limited variation that we observed in the duck/rabbit case, which neither the datum theory nor the thesis that percepts are constructed out of neutral material seem to be capable of handling.[22] Indeed, since a fully consistent datum theory requires, as we have seen, that the datum be itself unrecognized, there seems to be little important difference between the datum theory and the thesis that the mind constructs percepts out of a neutral material. Further, our approach is consistent with the way in which the recalcitrant data which lead to the overthrow of scientific theories appear. For while it recognizes the empiricist's point that a theory cannot be tested against observation if the observations are themselves completely determined by the theory, it also avoids the empiricist's embarrassment at attempting to explain how a meaningless datum can be relevant to the confirmation or disconfirmation of a scientific theory.

There is a second response to the claim that if observation is theory-laden there is no adequate reason for accepting one theory rather than another, for this conclusion does not follow at all from the analysis of perception developed in this chapter. Consider again our earlier example of two people passing by a steel mill, a child who only smells an odor of rotten eggs and a trained chemist who smells sulphur dioxide. Although the meaning of this odor is different for each of our perceivers and in each case depends on what the perceiver knows, it does not follow that the two percepts are on a par. It is clear that the chemist, who knows a great deal more about gases than the child, learns more as a result of his observation; his observation has greater epistemic value than that of the child, not because he has observed a pure datum, but exactly because there is more knowledge involved in his perception than in the child's.

Similarly, to borrow an example from Kuhn,[23] in the eighteenth century Lavoisier observed the properties of oxygen while Priestley, who still held the phlogiston theory, observed dephlogisticated air; the development of chemistry has given us good reasons for adopting Lavoisier's view. The thesis that the theory-ladenness of perception implies relativism is only plausible if one accepts the prior presupposition that only the observation of theory-free data can give us a reason for accepting one theory rather than another. If we do not accept this presupposition, then the rhetorical question, "Why choose one theory rather than another?" becomes a genuine question which must be dealt with by constructing an alternative analysis of theory change. This will be one of the tasks of the succeeding chapters, but before we can approach this question we must first reconsider the role that theories play in scientific research as a whole.[24]

Presuppositions

Normal Science   In 1687 Isaac Newton published in *Principia* what many take to
be the first comprehensive modern scientific theory. It consists
essentially of four propositions: the three laws of motion and the
inverse square principle of universal gravitation. From these four
propositions Newton deduced, among other phenomena, the laws
of planetary motion, the laws of falling bodies and of projectile
motion, and the variations of the tides. More important, Newton laid
the foundation of a way of thinking about physical realtiy which was
to dominate scientific research for more than two centuries. An
examination of some of the central episodes from the history of
Newtonian mechanics will serve as a basis for our discussion of the
role of theories in scientific research.

Late in his life Newton claimed that he had discovered the
inverse square law during the period from 1665-1667 and used it to
compute the force of gravity needed to keep the moon in its orbit,
and that the computed value and that derived from observation
"answer pretty nearly."[1] But Newton did not then publish his
results, for reasons that remain in dispute among historians. One
account, especially congenial to empiricists, asserts that Newton
did not publish his results because the agreement between
observation and theory was not sufficiently close. Reichenbach
writes:

> Newton himself saw clearly that the truth of his law depended
> on confirmation through a verification of its implications. In
> order to derive these implications, he had invented a new
> mathematical method, the differential calculus; but all the

brilliancy of this deductive achievement did not satisfy him. He wanted quantitative observational evidence and tested the implications through observations of the moon, whose monthly revolution constituted an instance of his law of gravitation. To his disappointment he found that the observational results disagreed with his calculations. Rather than set any theory, however beautiful, before the facts, Newton put the manuscript of his theory into his drawer. Some twenty years later, after new measurements of the circumference of the earth had been made by a French expedition, Newton saw that the figures on which he had based his test were false and that the improved figures agreed with his theoretical calculation. It was only after this test that he published his law.[2]

While the above explanation of Newton's failure to publish is plausible, historians have also proposed two others. One is that although Newton had found the agreement between theory and observation to be sufficiently accurate, he was faced with a theoretical difficulty. In his computations he had assumed that the earth and the moon could each be treated as if their entire mass were concentrated at a single point but he had not yet demonstrated that this assumption was justified.[3] The other interpretation is that Newton's late recollections of his youth were inaccurate and he had not yet developed the inverse square law.[4]

Whatever the correct explanation, Reichenbach's approach, which is typical of the empiricist view of science, should be evaluated in light of the fact that even after Newton did publish *Principia* there were major discrepancies between his theory and the results of observation and experiment. For example, he was aware that the computed value of the motion of the moon's orbit was only half the observed value. Having computed the motion of the moon's apsides, he simply stated that "The apse of the moon is about twice as swift"[5] and continued his exposition. Obviously Newton did not consider this discrepancy between theory and observation sufficient to withhold publication, but rather as a problem to be worked on. Eventually, around 1750, more than sixty years after the publication of *Principia*, the puzzle was resolved when Clairaut showed that the difficulty lay not with Newton's mechanics, but with the way in which the mathematics had been applied to the physical situation.[6] Similarly, to mention one more example, Newton's value for the speed of sound was off by twenty percent and this discrepancy between theory and observation remained unresolved for more than a century.[7] Both these problems were eventually resolved and the physicists' faith in the Newtonian system was justified. But according to traditional views of scientific method, physicists had no business acting in accordance with such a faith. The motion of the moon's apsides, for example, provided a clear counter-instance to the theory so that, if the empiricists are correct, the theory ought to have been rejected. What we find instead is that once Newton's mechanics had been accepted, difficulties such as the moon's motion and the speed of sound became research problems rather than counter-instances.

We have noted that one of the triumphs of Newtonian mechanics was the successful computation of the orbits of the planets, but this is not completely correct. Early in the nineteenth century astronomers recognized that none of the orbits for Uranus that had been computed on the basis on Newtonian mechanics fitted all the observed locations of the planet. Was this a counter-instance to Newtonian theory? If a counter-instance is an observation which leads to the immediate rejection of all or part of a previously accepted theory, then this observation was no more a counter-instance than was the observed motion of the moon's orbit; rather it became a research problem for astronomers. Within the framework of Newtonian mechanics there is one clearly acceptable factor which could affect the orbit of a planet: the existence of an as yet unknown planet exerting a gravitational force. Working independently, two astronomers, Leverrier and Adams, assumed that such a planet did exist and used the disparity between observation and theory as the basis for computing the mass and orbit of this planet. The planet Neptune was eventually discovered and Newtonian mechanics had achieved one of its greatest triumphs.[8] However, it was also known that there was a disparity between theory and observation in the case of the orbit of Mercury, and Leverrier used the same method to account for it, hypothesizing another new planet, Vulcan. Unfortunately no such planet exists and the orbit of Mercury never was accounted for within the framework of Newtonian mechanics.[9] Only with the advent of the general theory of relativity did an accurate computation of Mercury's orbit become possible, and only after the new theory replaced Newton's celestial mechanics did this failure of Newtonian theory come to be viewed as a counter-instance.

The above examples suggest a very different picture of the structure of scientific research from the traditional empiricist accounts. Rather than beginning from observed data and using it to confirm or reject proposed laws or theories, scientists such as Clairaut and Leverrier seem to have begun from an accepted scientific theory which guided their research and determined how they would deal with observed phenomena. As long as they worked within the confines of the theory, observational discoveries which, logically speaking, could have been taken as counter-instances became, instead, research problems to be solved by the proper application or further development of the theory. The kind of scientific research that these examples illustrate has been called "normal science" by Kuhn in order to distinguish it from "revolutionary" science,[10] scientific research which attempts to replace one accepted fundamental theory by another. Kuhn describes normal science as research done in accordance with a "paradigm," but exactly what he means by a paradigm has been a subject of wide debate, with one sympathetic critic claiming to have distinguished twenty-one senses of paradigm in his book.[11] It is not necessary for us to enter into this discussion here, for it is clear that fundamental theories that are presupposed by research

scientists form a major part of Kuhn's paradigms. It is the role of
such theories in scientific research that now concerns us and we
will use the term "normal science" to refer to research done in
accordance with an accepted theory.[12] Some further examples will
help to clarify the role of accepted theories in scientific research.

Astronomers in the ancient world believed that the earth was
stationary and the sun moved around it both in the course of a day
and in an annual rotation. In the third century B.C., however,
Aristarchus argued that the motion of the sun was merely apparent
and that in reality the earth rotated daily and revolved around the
sun in the course of a year. Let us focus on the annual revolution of
the earth. If Aristarchus' proposal is correct, then in the course of
half a year the earth moves a great distance to the opposite end of a
diameter of its orbit. There ought to be an observable shift in the
apparent location of the stars as a result of this motion. Aristarchus'
contemporaries put his proposal to this observational test, found
no parallax, and rejected it.[13] His suggestion that the stars are
much farther away than had previously been assumed, too far away
for the parallax to be observed, was not taken seriously, and
given the state of knowledge, with good reason. Aristarchus'
original proposal, in spite of its being *prima facie* implausible
to his contemporaries, was treated in the best scientific manner:
it was put to the available observational test and rejected; his
ensuing proposal must have appeared to his contemporaries as a
bit of *ad hoc* desperation.

Now let us consider the situation in, say, the eighteenth century,
when the moving earth hypothesis had been widely accepted by
astronomers along with a greatly expanded estimate of the size of
the universe. In this case too, the necessary parallax had not been
observed, but since the moving earth was the foundation of the new
astronomical tradition, the failure to observe the parallax did not
stand as a counter-instance, but rather as a problem to be solved.
Thus one of the central research problems of the time was the
attempt to construct a telescope which would permit the
observation of stellar parallax, a project which was finally brought
to completion by Bessel in 1838,[14] almost three centuries after the
publication of Copernicus' *De Revolutionibus*. Indeed, in the
sixteenth and seventeenth centuries the failure to observe the
stellar parallax was one of the standard objections against
Copernicus' version of the moving earth theory, although Galileo,
who accepted it, had already taken parallax to be a research
problem rather than a counter-instance.

> Now it might be said that there is a variation, but that it is not
> looked for; or that because of its smallness, or through lack of
> accurate instruments, it was not known by Copernicus. . . .
> Hence it would be a good thing to investigate with the greatest
> possible precision whether one could really observe such a
> variation as ought to be perceived in the fixed stars, assuming
> an annual motion of the earth.[15]

It is worth reflecting on what might have occurred had the ancient astronomers been able to observe the parallax entailed by Aristarchus' proposal. It is by no means clear that this would have led to the immediate rejection of geostatic astronomy, nor that a refusal to reject it would have been unscientific. The observed parallax might have become a research problem, just as the failure to observe the parallax along with the perturbations of the orbits of Uranus and Mercury became research problems for the modern astronomer.

Let us return to ancient astronomy for another example. One of its central principles, at least as important and far reaching as the geostatic principle, was that all the motions of the heavenly bodies are circular. Because of this principle the central problem of ancient astronomy was the problem of the planets. The planets, like all the other heavenly bodies, trace what appears to be an annual motion about the earth, but in the case of the planets this apparent motion is notably non-circular. Rather, the planets appear to move from west to east in an arc of a circle, stop and reverse direction for a bit (retrograde motion), then stop again and resume their forward motion, the net result being a kind of looped path. This motion would seem to present a clear counter-instance to the principle of circular motions, but it did not. Instead it became a problem to be solved, one which took up much of the history of astronomy.[16] Indeed, only when the motion of the planets came to be recognized as a problem did astronomy become a science. For centuries shepherds and others had observed the night sky and noted the retrograde motion of the planets, but having no prior beliefs as to how the heavenly bodies ought to move, they did not consider it to be especially puzzling or problematic. Only after Plato argued that all heavenly motions were really circular did the motions of the planets become problematic and scientific research on the problem begin. The principle that all heavenly motions are circular created a normal research tradition by creating a research problem, and also provided the criterion for determining an adequate solution to this problem: the motions of the planets had been accounted for only when they were shown to be the result of combinations of other motions which were themselves circular. From Plato until Kepler the principle that all heavenly motions are circular controlled astronomical research and any motions that deviated from this principle stood not as counter-instances, but rather as problems to be solved by resolution into circular motions.

Let us consider a final example, one from twentieth century physics. Early in the century it was discovered that the phenomenon of beta decay was inconsistent with the accepted principles of conservation of energy and momentum.[17] Again, logically speaking, this discovery could have been taken as a counter-instance and one or more of the conservation principles rejected. But the conservation principles are fundamental to the structure of modern physics and their rejection would have required a total reformulation of physics. This is not to suggest that we have some form of a priori knowledge of the truth of these principles or that they can never be reconsidered. But it does mean that they do not function as simple empirical propositions to be rejected at the first appearance of a counter-instance. Rather, they are protected propositions and any phenomenon that could logically be taken to be a counter-instance is interpreted as an apparent counter-instance, as a problem for which an explanation must be sought. In this case the explanation originally proposed by Pauli and developed by Fermi was to postulate the existence of a new particle, the neutrino, which simply had whatever energy and momentum was necessary to preserve the conservation principles. It was another twenty years before any independent evidence for the existence of the neutrino was obtained.[18]

All these cases illustrate the nature of normal science, in which an accepted fundamental theory serves to organize and structure scientific research. The theory determines the meaning of observed events by providing the scientist with grounds for recognizing which observations are relevant to his research, which of the relevant ones pose problems, how the problems are to be attacked, and what counts as an adequate solution to a problem. Now one of the traditional myths about the nature of science is that science is distinguished from all other intellectual activities, particularly philosophy, by its method. The scientific method, we are told, consists of suspending all our preconceptions and beginning research with a wholly unbiased search for facts. Galileo, the story goes, is the founder of modern science because, unlike the Aristotelians who turned to the texts of Aristotle for the answers to all questions, he carried out experiments and relied on his senses. But aside from the historical inaccuracy of this description of Galileo and his opponents,[19] let us consider on its own grounds what purely presuppositionless empirical research would be like.

Perhaps the only serious attempt actually to carry out scientific research in this manner is found in Francis Bacon's natural histories. Having attacked what he called the "idols"[20] that prevent the acquisition of knowledge, i.e., the prejudices and preconceptions which prevent us from discovering the facts, Bacon constructed a series of natural histories, compilations of all instances in which a given phenomenon appears, in order to provide a factual basis for the discovery of the laws of nature. His natural history of heat, for example, includes among its many instances such things as the warmth of the sun, the warmth of fresh animal excrement, and the warmth of hot herbs which are hot to the tongue although not to the hand.[21] What we have here is not an unbiased basis for knowledge, but a conglomeration useless until organized and decisions made as to which are indeed instances of the same phenomenon and which are worthy of further research. But such decisions must be made in accordance with some guiding principle, in the absence of which we have no way of knowing what data to collect nor what to do with it once collected. The point may be pressed home by considering what it would be like to make a list of, say, all the facts about one's room without any prior assumptions as to what facts are worth gathering. If one were to attempt such an investigation, noting every speck of dust on the walls and every scuff mark on the floor and making sufficient measurements to locate each feature with respect to every other feature, one would never get out of the room, indeed never finish tabulating every discoverable aspect of one square foot of the floor.[22]

In order to carry out meaningful research we require a research problem and some criteria of what evidence is relevant to its solution. More fundamentally, we require some basis for deciding what research problems are worth pursuing. It is our accepted theories, the systems of presuppositions to which we are already committed, which provide this basis. And because we always do research within a system of presuppositions, both our problems and our data are thoroughly theory-laden. As a result researchers always run the risks of pursuing dead-end problems and ignoring important data, but these are risks which must be taken if the pursuit of knowledge is to be possible at all. On the other hand, only those researchers who are thoroughly immersed in the thought patterns of a scientific tradition are subject to such happy "accidents" as the discovery of X-rays and of penicillin, since their presuppositions provide a set of expectations about an area of experience and thus make it possible for them to recognize some occurrences as being particularly significant. There are two theories of presuppositions from the history of philosophy that are particularly worth examining to help clarify the issues raised by the above discussion, those of Kant and Collingwood.

Paradigmatic
Propositions

According to Kant[23] knowledge is only possible in so far as we have experience, but we must distinguish between *experience* and *sensations*. This distinction, however, requires a prior distinction between *form* and *content*. Just as in logic we distinguish between the form of a proposition and its content, with both required to constitute a proposition, for Kant experience is also constituted of a combination of form and content, with sensation providing the content of experience while the mind itself provides the form. There are two ways in which the mind provides this form: via the forms of sense and via the concepts of the understanding. There are two forms of sense, space and time, and, for Kant, the fact that the objects we experience are found in space and time is a fact about the structure of the human mind and not a fact about the mind-independent structure of reality. And, exactly because space and time are contributed to experience by the mind, it is possible for us to know independently of any particular experience that all objects in the world that we experience through our senses will be located in space and time.[24] What we cannot know a priori is just what we will find in space and time when we look. This is the content of our experience, it is independent of the mind, and any knowledge of it can come only from actual experience.

Along with the forms of sense another faculty of the mind, the understanding, also plays a role in providing the form of our experience. The understanding has a small number of concepts, which Kant calls "categories," in terms of which sensations are organized and structured in the process of creating experience. The concept of causality will serve as an illustration. Since causality is a concept supplied to experience by the mind and thus part of the form of experience, we know a priori that every event has a cause, but we cannot know a priori what the particular cause of any given event is; this is part of the content of experience and must be discovered by empirical research.[25]

Each of the different aspects of the form of experience (space, time, causality and the remaining categories) supplies us with synthetic a priori knowledge, propositions which are known to be true a priori but which are nevertheless true of the experienced world, and it is these synthetic a priori propositions that serve as the presuppositions we are concerned with here. The role of space as a form of sense, for example, supplies us, according to Kant, with the knowledge that space conforms to Euclidean geometry. Similarly, the category of causality supplies the proposition, "Every event has a cause," and it is this proposition that we will consider further in order to illustrate the role of presuppositions in scientific research.

It has already been suggested that, for Kant, causality is constitutive of experience. Experience would not have the structure it has if it were not organized in accordance with the principle of causality, while causal connections are nonetheless a real part of the experienced world in spite of their being supplied to experience by the mind. But besides being constitutive of experience, causality is also, for Kant, constitutive of scientific research,[26] i.e., it serves as a presupposition which organizes and structures research. It does this in exactly the same way, although perhaps on a more fundamental level, as the presuppositions we have already considered, such as the principle of circular motion in ancient astronomy and the conservation principles in modern physics: by determining what are genuine scientific problems and by supplying the criteria for what is to count as a solution to such a problem. Thus for Kant the task of the scientist is to find the causes of events and a scientific problem is solved only when a cause for the event (or species of event) in question has been found. The synthetic a priori principle that every event has a cause guarantees that the search for causes must always be successful and thus that no event can count as a counter-instance to the causal principle no matter how long scientists have failed to find the specific cause. If scientists are unable to find the cause of an event, it is always the scientists and not the principle of causality that is at fault.

We have seen that for Kant the presuppositions of science are a priori propositions; they are necessary, eternal truths, i.e., there is no process by which they can be changed. For Collingwood, as for Kant, all knowing requires presuppositions, but they change throughout the course of human history. Collingwood's basic thesis[27] is that every meaningful proposition is the answer to some question and we can only understand the meaning of a proposition if we know what question it is intended to answer. But every question, in turn, has some proposition as its presupposition, i.e., something which I know or take to be the case. I cannot ask meaningful questions about phenomena of which I am totally ignorant. To ask, for example, what particle just passed through the cloud chamber presupposes that a particle just passed through the cloud chamber; to ask what the cause of a given event is presupposes that the event has a cause. Because questions have presuppositions which generate them, it is always possible to take one of two attitudes toward a question: we can answer it, or we can reject the presupposition and thereby reject the question. Those physicists, for example, who seek the causes of micro-events are accepting the Kantian presupposition that every event has a cause, while those who reject this presupposition with respect to micro-physics reject questions such as, "What caused this atom to decay at this moment?" as having no physical meaning. (We should note that it is also possible to misunderstand a question by attributing to it the wrong presupposition. If someone were to ask "Why did you buy a new car?" an answer such as, "Because I need transportation," would be acceptable in many circumstances, but it would be inappropriate if the questioner were a Freudian.)

It would seem that Collingwood has generated an infinite regress of questions and answers; he attempts to break this regress by drawing a distinction between "relative presuppositions" and "absolute presuppositions." The presuppositions we have been considering thus far, those which are propositions and thus themselves answers to questions, are all relative presuppositions. But at the root, so to speak, of each sequence of questions and answers, there is an absolute presupposition which is not the answer to any question, which is not, therefore, a proposition, and which is thus neither true nor false. Rather, an absolute presupposition is something like a methodological principle to be judged by what Collingwood calls its "logical efficacy,"[28] i.e., its fruitfulness in generating strings of questions and answers. Thus absolute presuppositions are the foundation of all intellectual activity; but, unlike Kant, for Collingwood what absolute presuppositions are held are characteristic of a given era and change in the course of history, although he gives no analysis of how change takes place beyond saying that absolute presuppositions develop "strains"[29] which lead to their collapse. In The Idea of Nature Collingwood maintains that each era in the history of science has been characterized by some fundamental conception of what nature is. If we may take these as examples of what Collingwood means by an absolute presupposition, he maintains that there have been three absolute presuppositions in the history of physical science: the Greek view of nature as permeated by mind, the early modern view of nature as mindless and acting in accordance with strict laws, and what he calls the "historical" view of contemporary science.[30]

There is much that is unclear or dubious in Collingwood's discussion of presuppositions. His thesis that absolute presuppositions are neither true nor false and that they are taken to be neither true nor false by those who propound them is most doubtful and the examples that he gives for approximately half of his *Essay on Metaphysics* do more to weaken his case than to help it. It is not clear whether each discipline has its own absolute presuppositions or whether they are characteristic of all thought in an era, nor even whether the answer to this question may not be different with respect to different eras in the history of thought. However, I am not concerned here to enter into an extended discussion of Collingwood's theory. What does interest me is his notion of changing presuppositions, because this notion, combined with Kant's notion of presuppositions which are constitutive of both research and of experience, will serve as the basis for a more adequate analysis of the nature of scientific presuppositions and of their role in research.

Perhaps the most striking feature of propositions which express presuppositions is that they do not fit into the customary dichotomy between analytic and empirical propositions. They are not analytic, since they are not formally tautologous and there is no sense in which the predicate constitutes a defining characteristic of the subject. If the principle that every event has a cause, for example, were analytic, then anything which could be shown not to have a cause simply would not be an event. But a working scientist who seeks the cause of some phenomenon does not have the option of deciding that he is not dealing with an event and thus eliminating his problem when he finds that he can discover no cause. Unless he is willing to question the causal principle itself, he must assume that further research will succeed in uncovering a cause. Similarly, a physicist who is working within the context of Newtonian dynamics must account for the motions of any material body by discovering the forces acting on it and the initial conditions. Now it is not uncommon for a philosopher to pick out one of Newton's laws and argue that it is analytic. The first law, it is often maintained, defines "inertial motion" or perhaps "uniform velocity" and the second law defines "mass" or maybe "force." But it is Newtonian theory *as a whole* that the scientist works with and if this "defines" anything, it is "material objects" since all material objects fall within its domain. Consider, then, a physicist who finds a case of non-inertial motion and cannot discover the forces which Newtonian theory tells him must be acting. What are his options? It is at least clear that he is not free to evade the problem by declaring the object in question non-material, which he could do if he were dealing with a definition. As long as he is working within the framework of Newtonian mechanics, he must assume that further research will turn up the forces that are responsible for the object's deviations from uniform straight line motion.

Let us consider the logical status of scientific presuppositions from another direction. When a scientist has difficulty finding phenomena that his presuppositions tell him must be present he will often carry out empirical research to seek these phenomena. This suggests two more arguments against the thesis that presuppositions can be expressed in analytic propositions. First, one of the central characteristics of an analytic proposition is that no counter-instance is logically possible. But, as we have just seen, counter-instances to Newtonian dynamics, for example, are logically possible and we know what could count as such: a body moving in some way other than a Euclidean straight line while no forces are acting on it, for example. By contrast, in the case of an analytic proposition such as "All bachelors are unmarried," we cannot specify what could count as a possible counter-instance. Second, a research project (e.g., Leverrier and Adams' computation of the orbit and mass of Neptune and the subsequent search for this planet) is often needed to eliminate a scientific scandal and show that the accepted presupposition is justified. Let us remember that we are not here dealing with synthetic a priori propositions. Although presuppositions are protected propositions which are not given up lightly at the first sign of a counter-instance, presuppositions do change and a persistent failure to account for an anomaly might well lead to such a change as in the case of the orbit of Mercury and Newtonian mechanics. Therefore it is often necessary to carry out empirical research in order to defend a presupposition, but such research is never necessary for the defense of an analytic proposition.

On the other hand such propositions are not ordinary empirical propositions, exactly because they are protected from straightforward empirical refutation. I will borrow a term from Kuhn and refer to propositions which express presuppositions and which are neither analytic, nor empirical in the usual sense, nor eternal truths, as *paradigmatic propositions*.[31] They constitute an epistemically distinct class in that they do not fit the traditional division of all proposition into a priori and empirical. Rather they are propositions which are accepted as a result of scientific experience but which come to have a constitutive role in the structure of scientific thought. At various times propositions such as that all celestial motions are circular, that physical space is Euclidean, that every event has a cause, or the entire panoply of modern conservation principles have achieved this status. Many of these have been taken to be necessary, eternal a priori truths, but some of them have nonetheless been abandoned, and it should by now be clear that every scientific proposition is subject to possible revision. This is the crucial lesson of Einstein's decision to abandon Euclidean geometry in favor of Riemannian. Before Einstein it had never been realized that we could respond to a refutation of a theory by leaving the physical postulates alone and changing the mathematics. From the point of view of logic alone, a counter-instance only tells us that there is something wrong in the accepted structure, it tells us nothing about where the problem lies.

It is considerations such as the above which led Duhem[32] a long time ago and Quine[33] more recently to maintain that we can never test an isolated scientific proposition, but only the entire body of science. This conclusion does not, however, square with what goes on in scientific practice where individual propositions are tested all the time. This is possible because even though there are no a priori propositions in science, not all propositions are treated as testable empirical hypotheses. It is only because a large body of knowledge is taken as paradigmatic that we can isolate individual propositions for purposes of testing, and what conclusions we draw from a particular test depends on what propositions are taken as paradigmatic. An example will illustrate this point.

By the middle of the nineteenth century it seemed to have been definitely established that light consisted of waves, not particles. Indeed, the experiments of Fizeau and Foucault appear to be classic examples of the use of crucial experiments to test single propositions. According to the wave theory of light, the velocity of light should be slower in a dense medium than in a less dense medium. According to the Newtonian version of the particle theory, the reverse should be the case. For some fifty years the demonstration that light had a greater velocity in a less dense medium was taken to be the final refutation of the particle theory. Nevertheless, Einstein introduced a new version of the particle theory in 1905 in order to account for the photoelectric effect. The experiments of Fizeau and Foucault still stood, they still provided a refutation of something in the Newtonian particle theory of light, only they did not refute the proposition that the experimenters thought they had refuted. Instead it was necessary to modify other hypotheses which were part of the nineteenth century version of the particle theory, hypotheses which the experimenters never thought of as the propositions under examination.

The above discussion provides us with a very different picture of the nature of scientific research from the traditional one which views the individual researcher as wielding the "scientific method" and adding his piece to the accumulating body of knowledge. We have seen that research is only possible after the scientist has committed himself to some theory, but we have also seen that the commitments of any scientist or of any research tradition may eventually be abandoned. The anomaly that the normal scientist seeks to explain may turn out to be a genuine counter-instance, and the phenomenon that is accepted as a counter-instance to a particular theory may have a perfectly acceptable explanation within that theory. But this is to say no more than that the search for truth involves the risk of error and that there is never any guarantee that we have made the right commitment. It may well turn out that the original commitment of an individual, or even of an entire tradition or culture, was a poor one and that whatever structure we have succeeded in erecting on it will collapse, but this is one of the risks that one takes in pursuing knowledge. To seek truth is to open oneself up to the possibility of error and the only way to avoid this possibility is not to inquire in the first place.[34]

Our discussion of scientific presuppositions has focused on their role in structuring scientific research; little has been said of how they structure the world of the scientist's experience. This question can best be approached in the context of another issue. A scientist who does research within the framework of an accepted theory has to learn his trade, this involves more than learning a set of paradigmatic propositions, and the trained scientist knows a good deal more than he could state as a set of propositions. The major part of his additional knowledge consists of knowing how to apply the theory to concrete problems. This is not learned by learning a set of rules of application in addition to the propositions which make up the theory. To learn a physical theory such as Newton's mechanics or Maxwell's electrodynamics and to learn how to apply it to specific problems is to learn one and the same thing. The student would seem to be trapped in a circle: on the one hand, he cannot apply the theory to the solution of physical problems until he understands the theory; on the other hand, he cannot understand the theory until he has learned to apply it to the solution of physical problems. But the circle is only an apparent one generated by presupposing a theory of mind which radically limits the powers it attributes to human thought, a theory which holds that intellectual activity must be in accordance with previously learned rules. A common example will help to clarify the kind of process that takes place in the learning of a scientific theory.

Consider a child learning to do arithmetic. Initially the child must memorize a set of rules: "two plus three equals five," "three plus five equals eight," and so forth. In the early stages of learning the child will be able to add any pair of two digit numbers and perhaps any pair of three or four digit numbers; but if asked to add a pair of, say, six digit numbers, he will reply that he has not yet learned to do that kind of problem. Somewhere in the learning process, however, the child reaches a stage at which he can add any pair of numbers of any length, even if he has never encountered numbers of this length in the classroom (and, of course, the same holds for columns of numbers of any length). It is only at this stage that we can say the child has learned to do addition. What occurs when the child acquires the ability to do any problem in addition? The immediate temptation is to say that he has learned a rule. But if this is the case, why not shorten the process by simply telling the child the rule to begin with and thus avoiding the tedious process of adding first one digit numbers, then two digit numbers, and so forth? The answer is that the child will not understand the rule and will be unable to use it. This "rule" is in fact an example of what Polanyi calls a "maxim": "rules, the correct application of which is part of the art which they govern."[35] The rule must be discovered by the learner for himself, and discovering the rule and learning to do addition are one and the same process.

Let us suppose that the general rule has been formulated. The student's ability to state the rule is neither necessary nor sufficient to do addition. One who can do addition knows considerably more than someone who can only state the rules, and his surplus knowledge cannot be passed on by stating a set of supplementary rules; the information can only be passed on to the learner by taking him through a course of problem solving which will lead him to discover the necessary techniques for himself. And as the child learns to handle numbers another phenomenon occurs; their meaning changes and he sees the symbols printed on the page in new ways. As the child progresses, learning to distinguish numbers from letters, odd from even numbers, prime from factorable numbers, and so on to the learning of various operations and even the theory of numbers, the meaning of numbers is continually enriched and what he sees when he looks at them on a page is continually transformed.

This is essentially the situation in which the physics student finds himself. Newton's laws, for example, can be stated briefly, but to memorize them is not to learn Newtonian physics. Indeed, the laws have no genuine meaning to someone who has not learned to apply them to the solution of dynamics problems. The student who, under the guidance of a teacher, practices the solution of problems, is learning a number of things: the meaning of Newtonian dynamics; a system of concepts and a language; a way of thinking and talking about the world; and a new way of seeing physical reality. To learn to deal with such common objects as automobiles and airplanes in terms of free body diagrams and D'Alembert forces is to learn to see them in a very different way from that of our everyday concern for transportation; the perceived world (or rather the relevant portion of the perceived world) takes on a new meaning for the physicist. Finally, the student is being initiated into a normal tradition: he is learning how a particular style of science is done, how one solves physical problems, and, eventually, what the unsolved problems are.

The discussion can be summed up by the notion of the "scientist's world:"[36] the system of meaning he perceives and in terms of which he carries on his research. As such, it is jointly constituted by the perceptual information he receives from the external world and by the theories to which he is committed. The physicist, attempting to understand the nature of reality, does so by the creation of theories, and the world he experiences is the result of the interaction of those theories and the reality that exists independently of our knowledge. Ideally the scientist would like simply to examine the structure of this independent world but, as we have seen in our discussions of the nature of perception and of the role of presuppositions in research, he has no direct access to it. His only access is through the creation of theories and the process of theory-directed research.

Only after the researcher has learned to see reality in terms of accepted theory is research possible, but it is also possible for the researcher to discover anomalies and thus come to reconsider accepted theories. Two factors operate here. First, theories often provide a definite description of what the scientist ought to see and thus sharpen his vision for the discovery of anomalies. Second, as long as the scientist is carrying on empirical investigation it is not theory alone which determines what will actually occur, but theory in conjunction with a theory-independent world. Whenever the structure of theory and the structure of the physical world fail to mesh, anomalies will appear and although many anomalous events may eventually be interpreted in terms of accepted theory, it is the recalcitrant anomalies that eventually lead to the overthrow of one theory and its replacement by another, i.e., to scientific revolutions.

Scientific Revolutions

The most interesting events in the history of science are revolutions: episodes, sometimes lasting decades, which result in the restructuring of the modes of thought of one or more disciplines and in some cases in the relationships between disciplines. On the deepest level two kinds of changes take place: both the presuppositions of a science and the concepts used in it are transformed and, as a result of these transformations, the world, or meaning structure, within which the scientist works and his research problems are also altered. The development and elaboration of these theses will be the main purpose of this chapter. We will proceed by means of a detailed analysis of two revolutions in physics: that initiated by Copernicus and brought to completion by Newton, and the more recent one brought about by Einstein's development of the theory of relativity.

The Copernican
Revolution

The revolution in physics initiated by Copernicus' development of a heliostatic astronomy is one of the richest and most thoroughgoing revolutions in the history of human thought. Restricting our attention to scientific issues, Copernicus' attempt to deal with an astronomical problem had the effect of under-mining the foundation of the accepted physics so that the new astronomy required the construction of a new physics.

For the medievals the physical universe was centered on the

earth and divided into two parts, the terrestrial sphere which consisted of the earth and everything below the moon, and the celestial sphere which contained the moon, sun, planets and stars. Each part was made of a different kind of material and had its own set of physical laws. There were thus two distinct systems of physics, terrestrial physics and celestial physics.

The terrestrial sphere was made up of the four elements: fire, air, water and earth. Each of the physical objects we normally encounter was taken to be a particular mixture of these elements. Each of the elements had its natural place, in which it tended to remain and to which it would return if permitted to move without constraint: earth downward toward the center of the universe, fire upward away from the center of the universe toward the sphere of the moon, and air and water to intermediate positions with air normally higher than water. The center of the universe is not defined as the center of the earth; the two coincide because the center of the universe is the natural place towards which the earth tends to move, although it is in principle possible for a sufficiently strong force to remove the earth from the center of the universe.[1]

The concept of motion to a natural place leads directly to a distinction between natural motion and violent motion. Any motion which opposes the natural motion of an object, such as lifting an earthy element, is violent, and requires an external force. As soon as the external force is removed, natural motion ensues and the object returns to its natural place. Natural motion, then, is always of finite duration. This analysis of motion was capable of providing explanations for various "observed" phenomena: it could explain why heavy bodies fall and flames rise, why the oceans rest on the earth and the air remains above the oceans, and why the earth is at the center of the universe.

However, it also left a number of unexplained phenomena and thus a number of research problems. The most important was the problem of projectile motion, which dynamicists worked on from the days of Aristotle until it was finally eliminated by the new physics—eliminated, not, as we shall see, solved in the sense in which, say, Leverrier and Adams solved the problem of the perturbations of the orbit of Uranus. The problem is essentially this: consider a projectile such as an arrow shot from a bow. After it leaves the bowstring the arrow continues to move forward for some distance, but eventually it lands and comes to rest. Since the arrow is an earthy object its natural motion is vertically downward and its horizontal motion is violent motion, but all violent motion requires some external force to maintain it and in the absence of such force the arrow leaving the bowstring should drop directly to the ground. The problem, then, is to find the force which accounts for the violent motion of the arrow (much as Leverrier and Adams had to find the force which could account for the perturbations of Uranus and of Mercury). Among the attempted solutions were Aristotle's proposal that the air supplied the force which kept the arrow moving[2] and the medieval theory that the bowstring imparted a force or "impetus" to the arrow. The point to be emphasized here is that the attempt to account for projectile motion was a genuine research problem for ancient and medieval dynamics.

In the second realm of ancient physics, the heavens, none of the four terrestrial elements is to be found. The stars, planets, sun and moon are made up of a different and more perfect element called "quintessence" or "ether," the natural motion of which is circular. The reasons for the choice of circular motion would seem to have been partly observational and partly theoretical (some would say "religious" but it is far from certain that we can make the sharp distinction between scientific and religious ideas for the Greeks—or even for Newton—that we would make today). The daily motions of the heavens and many of the annual motions appear to be circular, and the celestial bodies were believed to be perfect, unchanging objects. But motion was taken to be a form of change; hence if the celestial bodies moved their proper form of motion must be the one which is most nearly unchanging, an eternal circular motion in a permanent orbit. Indeed, strictly speaking it was not the celestial bodies that moved; they were attached to spheres which rotated in place. Unfortunately, not all the observable motions of the celestial bodies are circular: the planets[3] a small but prominent group of observable objects, have annual motions which consist of strange looped paths.[4] This exception to circular motion provided the main research problem of astronomy from Plato to Kepler.

Two comparisons between the problem of the planets and that of projectile motion require special emphasis. First, we are dealing with two different bodies of dynamical presuppositions; the kinds of events that require explanation differ in the two cases. In the case of the projectile, departures from *vertical* motion require an account; for the planets, departures from *circular* motion require an account. As we shall see, after Newton brought celestial and terrestrial motions within the scope of a single theory, the same kinds of motion required explanation in all cases. Second, the kinds of explanation that the operative presuppositions permit in the terrestrial and astronomical cases are different. The concept of violent motion applies only within the terrestrial sphere. While it is possible to admit a departure from natural motion there and thus to seek a force to account for the departure, no violation of natural motion is permitted in astronomy. The planets, it was maintained, do indeed move only in circles, in spite of the fact that they appear to move in non-circular paths, and the astronomer's research problem was to find a system of circular motions which will "save the appearances," i.e., explain why the planets appear to us to move in non-circular paths.

It is in this intellectual context that the reception of the Copernican hypothesis must be understood. We have often been told that the Copernican system is simpler and more accurate than the old earth-centered astronomy with its circles and epicycles, and from a purely formal point of view there is some substance to this claim. By taking the sun to be stationary Copernicus eliminated the major epicycles and the need for *ad hoc* hypotheses to explain the fact that Mercury and Venus never move far from the sun, and he was able to determine which of these two planets was closer to the sun. On the other hand, his system was no more accurate than the older theory, and he also retained the principle of circular motion for celestial bodies and thus required epicycles. Indeed, he even retained the notion that the celestial bodies are carried by crystalline spheres, to which the title of his book, *On the Revolutions of the Heavenly Spheres (De Revolutionibus Orbium · Coelestium),* refers.[5]

To claim, therefore, that the Copernican view was simpler *simpliciter* is to view the question ahistorically, from the viewpoint of contemporary science rather than in the intellectual context in which it emerged. Setting aside questions of the interrelation between science and theology and confining ourselves to scientific questions in the modern sense, the price to be paid for the minimal formal gains of the new theory was the undermining of the accepted physics, while, at least initially, the Copernicans had no new physics with which to replace it. By putting the earth into orbit around the sun Copernicus destroyed the accepted distinction between the sublunar and superlunar spheres and, in effect, made the earth a celestial body. It might still be possible to take circular motion as the natural form of motion for celestial bodies, but the concept of a celestial body was altered and the entire system of terrestrial mechanics undercut. The earth, the paradigm case of an "earthy" body, no longer rested at the center of the universe; how then could we account for the fact that other earthy bodies tended to fall to the earth? Furthermore, problems, such as why the earth did not leave its atmosphere behind, and why a stone thrown up fell straight down, arose. If the air and a stone had a natural vertical motion and the earth were moving about the sun, it would require an additional violent motion to keep them moving along with the earth; but no agent of violent motion appeared to exist. The new astronomy would require a wholly new form of dynamics, one which would dispense with the notion that different dynamical laws applied to the heavens and the earth. Copernicus did not supply one, and only after a new dynamics had been constructed could the revolution he initiated be completed.

The first major step toward a new physics was taken by Galileo with the introduction of the concept of inertial motion: if a terrestrial body were in motion with no force acting on it, it would continue in its motion indefinitely. This thesis, once accepted, eliminates the old problem of projectile motion. The idea that objects have a natural motion remains, but in the case of terrestrial objects natural motion is no longer finite but indefinite, and it becomes necessary to explain why motion stops rather than why it continues. In the case of projectile motion the old problem of why the arrow continues to move after it leaves the bowstring is dissolved: this is the natural thing for it to do; no explanation is required. Thus Galileo proposed much more than a new theory. He offered what Toulmin has called a new "ideal of natural order,"[6] a new fundamental conception of how nature acts, and as a result he changed our understanding of what phenomena require explanation and of what questions can legitimately be asked.

But Galileo did not complete the job of formulating the new dynamics for, although he introduced a major part of the modern concept of inertia, he took inertial motion to be circular. If a projectile were shot from the earth, for example, and no other forces interfered, it would continue to move eternally around the earth in a circular path. Galileo was a contemporary and correspondent of Kepler but he seems to have simply ignored his discovery that the planets move in ellipses, not in circles. On the other hand, by maintaining that circular motion is the natural form of motion for terrestrial bodies, Galileo took a long step in the process of breaking down the time-honored distinction between terrestrial and celestial bodies. It was left, however, for Descartes and Newton to take the final step and propose that inertial motion, for both celestial and terrestrial bodies, is straight line motion. As a result of this new concept of inertia the physicist's conception of nature changed once again, and along with it his understanding of which kinds of phenomena require explanation, and the standards for an explanation, also changed. Motion in a straight line at constant velocity required no further explanation than to say that "no forces are acting," which, in the context of Newtonian mechanics, is not an explanation at all but a denial of the need for one, whereas circular or any other non-linear motions became the deviations that required explanation. Most important, the same kinds of problems arose and the same kinds of solution were needed for both terrestrial and heavenly bodies.

Thus, this scientific revolution involved fundamental changes in the presuppositions of physics, along with changes in some of the basic concepts in terms of which scientists thought about the physical world. We have concentrated thus far on changes in presuppositions; we turn now to some of the conceptual changes.

Conceptual
Change

Both before and after Copernicus astronomers were challenged by the strange motions of the planets and we have seen that they explained them differently before Copernicus and after Newton. But while this change in scientific presuppositions was in progress the concept of a planet was itself altered, so that in an important sense the new astronomers were not attempting to explain the same thing as the old astronomers. The point can be made most clearly by examining the changes that occurred in terms of the distinction between the sense and reference of a concept, both of which were altered in the present case.

Taking the sense of the concept first, before Copernicus the defining characteristics of a planet included the requirements that it move around the earth and in relation to the fixed stars. For Kepler and Newton motion around the sun had become a defining characteristic. Similarly, later developments in astronomy led to further changes in the concept of a planet so that it now makes perfectly good sense to suggest that there are planets orbiting stars, a claim that would have been meaningless to a pre-Copernican and even for Copernicus himself.

The thesis that the meaning of scientific concepts changes as a result of a scientific revolution has been regarded by many logical empiricists as one of the most outrageous claims of the new philosophy of science. It has long been a central doctrine of empiricist thought that the meanings of terms are completely independent of the propositions in which they occur and that we can accept or reject propositions without this having any effect on what we mean by the terms which occur in them. One of the implications of our view, however, is that there is an intimate relation between the content of concepts and the propositions in which they occur. In defending against this view one logical empiricist, Israel Scheffler, concedes that the sense of concepts changes in the course of the development of science, but maintains that this is unimportant since, "for the purposes of mathematics and science, it is sameness of reference that is of interest rather than synonymy. . . ,"[7] i.e., no significant conceptual change takes place unless the extension of the concept has changed. But if we return to the concept of a planet it is clear that the extension of the concept also changed. Before Copernicus the earth was not included in the class of planets; for later astronomers the earth had become a planet. Similarly, the moon and sun were planets for the pre-Copernicans since they moved around the earth and relative to the fixed stars,[8] but they were no longer planets after Copernicus. Indeed, a new concept had to be introduced to accommodate the satellites of planets.

Let us take a second, more complex example. It has already been intimated in our discussion of the concept of a planet that we cannot fully understand the way a concept functions by examining it in isolation from other concepts. A full analysis of the transformation of the concept of a planet from medieval astronomy to the present requires an analysis of the changes in such related concepts as a sun and a star. Again both the sense and reference have been changed. For the medievals the sun was a unique object and it would have made no sense to call the sun a star or to call a star a sun; for the modern astronomer the terms "star" and "sun" are synonymous. (They are not synonymous in everyday language, but our concern is with the astronomer's technical vocabulary.) As for the reference of these concepts, for us there exists one more star and billions more suns than the pre-Copernican's could, in principle, have admitted.

Consider now another pair of closely related concepts, fall and heaviness or weight. Ancients, medievals and contemporaries all assert that heavy bodies fall, and offer explanations for this phenomenon, but not only do the explanations differ, the sentence "Heavy bodies fall" has different meanings in the contexts of Aristotelian and of modern science because of differences in the meanings of the terms "weight" and "fall." For Aristotle there is not only a relative distinction between "light" and "heavy" but an absolute distinction: lightness is as much a real property as is weight;[9] heavy objects move downward to their natural place, light objects upward to theirs. Aristotle's concept of a heavy body differs from ours since it is tied to a contrary property which does not exist in the modern conceptual scheme. Similarly, his concept of fall is unlike ours since not only is its sense different—motion to a natural place for Aristotle, motion to a gravitating body for us—but its extension as well. For Aristotle a stone moving to the ground and the upward motion of a spark or a helium filled balloon are all instances of the same kind of motion.

The Aristotelian concept of fall as motion to a natural place did not give way at once to the Newtonian concept of fall as motion under the influence of gravitation. Fruitful concepts evolve as they are incorporated into new theories, sometimes undergoing many changes in the process. For example, Galileo, who is so often credited with having discovered the modern form of the law of falling bodies, introduced a crucial change into the Aristotelian concept of fall and thereby produced a concept quite different from either the Aristotelian or the Newtonian—but one much closer to the former. Galileo accepted the concept of fall as motion to a natural place, but rejected Aristotle's analysis of natural places For Aristotle, as we have seen, space is structured independently of matter, and up and down are inherent properties of space.[10] For Galileo, on the other hand, the natural place of a material body is determined by its source. The natural place of a piece of the earth, for example, is the earth, therefore a stone returns to the earth when dropped; a piece of material from the moon, if left unconstrained, would fall back to the moon.

> Now just as all the parts of the earth mutually cooperate to form its whole, from which it follows that they have equal tendencies to come together in order to unite in the best possible way and adapt themselves by taking a spherical shape, why may we not believe that the sun, moon and other world bodies are also round in shape merely by a concordant instinct and natural tendency of all their component parts? If at any time one of these parts were forcibly separated from the whole, is it not reasonable to believe that it would return spontaneously and by natural tendency?[11]

This change in the notion of a natural place eliminates one of the standard problems raised by the Aristotelians against the Copernican view, the problem of why earthy objects fall downwards to their natural place at the center of the universe if the earth itself does not fall there. By redefining the natural place of an earthy object as the earth itself, wherever in space it might happen to be, Galileo can view the fall of a stone as a return to its natural place while eliminating any inconsistency between this claim and the Copernican claim that the earth is in motion around the sun. With Newton the concepts of natural place and of tendencies to return to the whole are done away with and the concept of fall undergoes another transformation, into motion under the influence of gravitational force. Later, in the context of general relativity with its banishment of forces and introduction of motion along geodesics in space-time, the concept has undergone yet another transformation, and we have no sure reason for believing this latest transformation to be the final one. Besides concepts which change in the course of a scientific revolution, there are others which are dropped altogether. This has happened with natural place, phlogiston, and ether, and it is at least logically possible that some future development in physical theory will eliminate the concept of fall. But the situations which concern us at present are those in which a concept is transformed. Indeed, as the fate of absolute lightness and natural place shows, the abandonment of some concepts may result in modification of those retained.

Cases of transformed concepts provide sameness/difference situations analogous to those we have examined in the case of perception: after a revolution we find recognizable modifications of old concepts. In the concept of a planet, for example, both the sense and the reference changed, but many aspects of both remained the same throughout this change. For example, planets are still celestial objects which move with respect to the fixed stars, and there is considerable overlap between the extension of the old concept and of the new one.[12] Similarly, in spite of fundamental changes, the concept of fall still applies to the motion of a wide class of objects left unsupported near the surface of the earth. We can clarify this point and advance our analysis of concepts and their relation to propositions in scientific theories by considering another example, this time from Newtonian mechanics.

How are we to define the Newtonian concept of mass? Newton gives two definitions. In the first definition in *Principia* he defines "mass" as the quantity of matter to be measured by the product of density and volume,[13] while in the third he defines "inertia" as the power of a body to resist changes in its state of motion and notes that mass and inertia are two different ways of conceiving the same property.[14] While contemporary textbooks avoid such verbal definitions, writers do differ fundamentally in the ways in which they introduce the concept. Some, for example, take "mass" to be an undefined primitive;[15] others view "force" as a term that we understand from everyday experience and use Newton's second law to define "mass" as the ratio of force to acceleration.[16] But none of these definitions by itself tells us how the concept of mass functions in Newtonian mechanics—although the possibility of alternative definitions does suggest that there is an intimate relation between the concepts of mass, force, and acceleration and that it may be impossible to understand any of these concepts without understanding all of them. We must examine, then, how the concept of mass functions in the structure of Newtonian physics.

Mass appears in the two central equations of Newtonian mechanics, the second law, $F=ma$, and the gravitation formula, $F=Gmm'/r^2$. Learning the Newtonian concept of mass and learning to use these equations to solve physical problems are inseparable from each other, so no clear distinction can be drawn between understanding the concept of mass and understanding the fundamental propositions in which it occurs. Further, Newtonian mechanics is often formulated in terms of the vector calculus. In these terms, force, acceleration and distance are vector quantities and mass is a scalar, so that a full understanding of the concept of mass requires understanding the distinction between a vector and a scalar and how to manipulate vector equations. (Most contemporary physics students learn to manipulate vectors in the course of learning Newtonian mechanics.) To take one more example, Newtonian mechanics makes extensive use of differential and integral calculus. In these terms the following two equations are equivalent: $F = d(mv)/dt$ and $F = mdv/dt$. Understanding why these equations are equivalent and why we cannot also write $F = vdm/dt$ is part of what is involved in understanding the concept of mass (as well as the Newtonian concepts of force, acceleration, velocity, momentum and time). We cannot learn the concepts of Newtonian physics in isolation from each other, nor in isolation from the propositions and formulas in which they appear, nor, since Newtonian physics is mathematical physics, in isolation from the mathematical operations that are accepted as legitimate. To learn Newtonian physics is not first to learn concepts and then put them together into propositions; it is simultaneously to learn a constellation of concepts and the propositions and formulas in which they occur.

A metaphor used by some logical empiricists can be adapted here to further clarify the relation between concepts and propositions. In the course of a critique of operationism Hempel writes:

> Scientific systematization requires the establishment of diverse connections, by laws or theoretical principles, between different aspects of the empirical world, which are characterized by scientific concepts. Thus the concepts of science are the knots in a network of systematic interrelationships in which the laws and theoretical principles form the threads.[17]

Since writing this passage Hempel's views on the problem of meaning have continued to develop. As we have seen, he now fully recognized that the meanings of theoretical terms cannot be completely specified by reference to an antecedently available vocabulary, but he concludes from this that the problem of the meaning of theoretical terms "does not exist."[18] Hempel's conclusion should be that this problem as it has been conceived by logical empiricists has no resolution, but, because he retains his empiricist commitment, Hempel fails to recognize that he has suggested an alternative approach to the theory of meaning, one which will provide much insight into the problem of conceptual change.

A scientific concept is a knot in a web; the strands in the web are the propositions that make up a theory; the meaning of a concept is its location in the web. Thus the meaning of a concept is determined by the strands that come into this knot, by the other knots the one in question is connected to, and by the further connections of these other knots. In the case of the concept of mass, the second law and the principle of gravitation are two of the major strands, and this knot is also tied to the knots which constitute the concepts of force, acceleration, etc. But the distinction between scalars and vectors and the differential and integral calculus also contribute strands to this knot. In short, a concept is not a simple which is grasped entirely or not at all, but rather a complex which can be learned a bit at a time. It is common for a student to learn to use such concepts as force, mass and momentum and a law such as the $F=ma$ form of the second law before he has studied calculus. When he later learns calculus and can handle the momentum formulation of the second law, his understanding of this law, as well as of these concepts, is altered. Bit by bit as he develops his understanding of a theory, learning more of the strands that make up the web, he also develops a fuller understanding of the concepts involved.

Earlier I argued that Feigl was wrong in holding that a theory is a meaningless intellectual construct until it is connected up to observations; I suggested that rather than observations giving meaning to the theory, it is more nearly correct to hold that theories give meaning to our observations. Now my position can be further clarified. The picture of a scientific theory as a system of propositions and concepts existing independently of any connection with observation does not describe a stage in the historical development of any actual theory, nor does it throw light on the structure of scientific theories.[19] Just as there can be no significant observation without theories, so there can be no scientific theory that is not used to organize some area of experience. Thus there is a sense in which it is true that observation confers meaning on theories, for part of understanding a theory is understanding what areas of experience it applies to and relevant observations provide an important set of threads in the theoretical web; at the same time, the theory supplies the meaning of the observations. Concepts, propositions and observations are the elements from which scientific theories are built. It is impossible to introduce any one of these elements without introducing the other two, and impossible to begin learning one of these aspects of a theory without beginning to learn the others.

Finally, we can apply the web metaphor to concepts that are transformed as a result of a scientific revolution. Some of the strands which come into a particular knot are removed, others are redirected, and some new strands are introduced. The concept retains some of its old characteristics since some of the old strands are left intact, but it also loses some old relations and acquires new ones, thus we acquire a new version of an old concept.[20]

These themes can be further developed and our arguments

reinforced if we turn to the transformation from Newtonian to relativistic mechanics.

The relation between relativity theory and Newtonian mechanics has become a subject of extensive debate, the central issue being whether the development of relativity theory constitutes a scientific revolution: a transformation in the fundamental presuppositions of physics along with concomitant changes in scientific concepts and the scientific world. Logical empiricists and most physicists deny that such a revolution has taken place. Relativity theory, they maintain, has not superseded, but rather is a generalization of, Newtonian mechanics to include high velocity situations. The equations of Newtonian mechanics had been verified for low velocity situations only, it is argued, and the assumption that they applied to high velocities as well was gratuitous; Einstein produced a new set of equations which apply to both high and low velocities and which reduce to the Newtonian equations in the limit of low velocities. In response to this view writers such as Kuhn[21] and Feyerabend[22] maintain that although Einstein's equations can be shown to reduce to equations which are formally identical with Newton's for velocities that are small in comparison with the velocity of light, they are not Newton's equations since the key terms are defined differently. Let us examine these issues more carefully.

It should be clear that formal similarity alone is not sufficient to prove that two equations are identical; they may well be about entirely different subject matter. Ohm's law, $V=IR$, for example, is formally identical to $F=ma$ and to $s=vt$, the formula for computing distance from velocity and time, but no one would suggest that these equations all say the same thing. Similarly, the differential equation for a circuit with resistance, inductance and capacitance is formally identical with the equation for a forced vibration, and so forth. These formal analogies provide the basis for a large number of computation techniques and for analog computers, but they do not show that there is any significant sense in which the equations involved are identical; for two equations to be identical they must not only have the same form, but their symbols must stand for the same concepts. Let us consider, then, the relation between the concepts of Newtonian and relativistic physics.

Continuing to take mass as our main example, we have already seen that for Newtonian mechanics the mass of a body is a constant; for relativity theory the mass of a body is a variable dependent on velocity, with a minimum value equal to the mass as given in Newtonian theory and an upper value which increases without limit as the relative velocity of a body approaches the velocity of light. As a result of this alteration, many of the patterns in which the concept of mass appears in relativity physics are different from those in which it appears in Newtonian mechanics. The formula F=ma no longer holds, nor could we define m as a proportionality constant between force and acceleration. The formula F=d (mv)/dt does still hold but, since mass is now a variable which is dependent on velocity and velocity is a function of time, it is no longer permissible to pull the m out of the derivative. Rather, the momentum, mv, becomes fundamental and the formula F=d (mv)/dt is customarily used as a definition of "force." But the concept of momentum in which m appears does not function in the same way as does the concept of momentum in Newtonian mechanics. Even though momentum is still defined as mv, the actual use of this notion inserts m into further new conceptual patterns. Thus Newtonian mechanics has two distinct fundamental principles, the conservation of momentum and the conservation of energy, with m also appearing in certain characteristic patterns in the energy equations. But relativity theory replaces these separate principles with a single combined principle of conservation of momentum-energy, and the concepts of momentum and of energy are no longer related as they were in Newtonian mechanics. The new relation between momentum and energy indicates that the concepts of momentum and of energy have been changed, and concepts such as mass, which appear in the new equations in new patterns and subject to new operations, share in this conceptual change. Let us take one more example.

In Newtonian mechanics the concepts of mass and of energy are, to some degree, related since, for example, kinetic energy is a function of mass. But in relativity theory the tie is much more intimate: the mass of a body at rest is identical with a form of energy and the kinetic energy of a moving body is identical with an increase in its mass. Thus the Newtonian equation $T = \frac{1}{2}mv^2$ and the relativistic equation $E = mc^2$ are different kinds of equations, i.e., the equals sign asserts something radically different in each. The former equation tells us how to compute the numerical value of the kinetic energy of a moving body from the numerical values of the body's mass and velocity; the latter equation tells us how to compute the energy of a body, but it applies to both moving and stationary bodies and it asserts the equivalence of energy and mass. The difference is akin to the difference between asserting (a) "The morning star has the same magnitude as the evening star" and (b) "The morning star is the evening star" (where, of course, (b) entails (a)). But we cannot recognize the difference between $T = \frac{1}{2}mv^2$ and $E = mc^2$ from an examination of the equations alone; this requires an understanding of these equations in the context of the theories in which they function. We must, for example, understand the special role that the velocity of light plays in the theory of relativity in order to be able to understand why $E = mc^2$ can assert the identity of E and m and not the identity of E and c. We can only grasp the meaning of the symbols if we understand the concepts they stand for, and we can only understand these concepts if we understand the role that they play in the theory.

Let us not, however, forget the other side of the coin. For while the shift from Newtonian to relativistic physics has resulted in fundamental changes in a number of physical concepts, these concepts have also remained the same in many respects. Mass, for example, remains a measure of resistance to change of velocity, although the way in which velocity changes has been reformulated, as well as playing a role in the concepts of energy and momentum. Similarly, the concept of time is fundamentally altered by relativity theory: it too becomes a function of relative velocity. Time measurements vary for different frames of reference, the concept of two events being simultaneous while at a distance from each other vanishes, and under certain conditions gaps between events that are temporal in one frame of reference are spatial in another. Nonetheless, the relativistic concept of time is a modification and development of the Newtonian concept of time (but not a generalization of it) and it does many of the same kinds of jobs that the concept of time does in Newtonian mechanics, e.g., in the definitions of velocity and acceleration. Thus the concepts of relativity theory are, in an important sense, continuous with those of Newtonian mechanics while being nonetheless different concepts. As a result the formulas that the two theories seem to share are different formulas: understood in the context of their respective theories, they say different things about the structure of the physical world.

There is, however, a second reason why the thesis that relativity theory is a generalization of Newtonian mechanics remains attractive. Newtonian mechanics gives completely satisfactory quantitative results, indistinguishable from those of relativity theory within the limits of observational precision, for low velocity situations. But let us consider the status of equal numerical results more closely.

Since Newtonian mechanics gives numerical results which are consistent with observation for a wide range of cases, any new theory which is to apply to the same phenomena must yield results that are at least very close to those derived from Newtonian theory. Now the fact that relativity theory can be made to yield equations which are formally identical to the Newtonian equations for just that class of situations to which Newtonian mechanics can be successfully applied is a particularly convenient way of guaranteeing that the two theories will give the same numerical results,[23] but this does not entail that we are dealing with a single theory. Rather, we have two different theories which give the same numerical results for a particular class of situations and different results for other situations.

An examination of the way in which numbers are manipulated in scientific practice should make us wary of drawing inferences from the identity of the numerical results to the identity of theories. Any number that is derived from a theory has a meaning attached to it which is supplied by that theory, and any number that is derived from measurement has a meaning that derives from the theories on which the construction of the instruments used is

based. In some cases the difference in units is enough to make
this point clear, since it is obvious that "five dynes" is not the
same thing as "five feet," but there are also cases in which two
numbers with the same units have very different meanings. "Five
amperes," for example, could in one case denote a steady state
current and in another a transient current at some time t. "Five
electron volts" could denote the potential difference between an
anode and a cathode or it could denote the rest mass of a particle.
Thus we must know the meaning attached to a number before we
can make sense of it.

This might appear trivial until we recognize the ease and
frequency with which the meaning of a particular number may be
ignored as the number is shifted from one context to another in
which it has a completely different meaning. This is particularly
clear in the case of analogies used for purposes of computation.
Consider, for example, an extremely simple case of an analog
computer: an electric circuit consisting of a variable pure resistance, a variable source of voltage, and an ammeter. According
to Ohm's law the current equals the voltage divided by the resistance, $i=V/R$, and for motion at a constant velocity, the velocity
equals the distance divided by the time, $v=s/t$. Since the two
equations have the same form, in any case in which the same
numbers are plugged into both equations in the corresponding
places the equations will yield the same numerical result; the
meaning associated with these numbers plays no role in the
computation. Thus I can use the circuit as a means of computing
velocities since if I have numerical values for the distance and
the time I can ignore the meaning associated with these numbers,
adjust the voltage and the resistance, and read the appropriate
value for the velocity off the ammeter. To be sure, the result that
I read off the ammeter is "i amperes," but given the formal similarity between the two cases, I know that i is numerically equal to
v and I thus ignore the significance of i itself and replace it with
that of v.

A similar approach can be used to simplify computations in
many cases, sometimes without the use of hardware as an
intermediary. The structural engineer, for example, is often faced
with the problem of analyzing a statically indeterminate structure,
i.e., a structure which has more reaction components to be
determined than can be found by using the equations of statics.
There is a general method for solving these problems but it is
extremely cumbersome and a number of shortcuts have been
developed. One such shortcut is the "column analogy," which can
be used in the analysis of frame structures. A general analysis
shows that the equations for the redundant reaction components
(i.e., those in excess of the number which can be found from
statics) is formally similar to the equations for the stresses on a
statically determinate column whose cross section is of the same
shape as the frame. Thus it is possible to describe an imaginary
column, calculate the appropriate stresses, and be sure that these
are numerically equal to the desired reaction components on the

frame. The engineer gets the correct numerical result but the meaning of the number is changed as it is transferred from one context to another.

In a completely analogous way, the physicist who uses classical mechanics for dealing with a low velocity situation is not making use of a special or limiting case of relativistic physics, but is taking advantage of a formal analogy in order to simplify computations. Let us remember that the use of Newtonian mechanics does constitute a simplification, for it is not literally true that the equations of relativity theory "reduce" to those of classical mechanics for the case of a slowly moving body. The reduction is only complete in the case in which the velocity of the body in question is zero. For dynamical problems Newtonian and relativistic mechanics never give the same result. The justification for using Newtonian mechanics in lieu of relativity is that for a certain range of cases and a permissible margin of error, the difference between the quantitative results supplied by the two theories can be ignored so that we might as well use the simpler equations of classical mechanics. This, as Kuhn had pointed out,[24] is comparable to the justification for surveyors using an earth centered astronomy. The technique used provides a numerical result with a particular meaning but, since the scientist has independent reasons for accepting this numerical result, he ignores the meaning associated with the theory actually used and replaces it with the meaning provided by the presently accepted theory.

That classical and relativistic mechanics never give exactly the same numerical result for any moving body suggests another line of analysis. Still taking mass as our example, for Einstein $m = m_0/\sqrt{1-v^2/c^2}$ where m_0 is numerically equal to the Newtonian mass. In using this equation the physicist will often expand it as an infinite series, $m = m_0[1+\frac{1}{2}(v/c)^2+\frac{3}{8}(v/c)^4+\frac{5}{16}(v/c)^6+\ldots]$, and the claim that the relativistic mass reduces to the Newtonian mass for low velocities can be expressed as follows: for values of v/c which are sufficiently small, all terms in the brackets after the first are so small that we can ignore them and take $m=m$. Now let us make this notion of a sufficiently small velocity precise by supposing that we can ignore any set of terms in the above expansion, provided that doing so will introduce an error of not more than 10%. A straightforward computation shows that ignoring all terms other than the first is then legitimate for velocities up to $v=.45c$. Now the notion of a small velocity is, of course, a context-relative notion. In everyday terms, and even in space age terms, a velocity of .45c, or some 83,000 miles per second, is enormous, but it is not especially large in the experiments of particle physicists, which involve velocities well over .9c. It might be said, then, that for any velocity less than .45c the relativistic mass "reduces" to the Newtonian mass and relativity theory "reduces" to Newtonian theory. If we now calculate the value of v for which including the second term of the series but ignoring all further terms will yield an error of 10%, we get a result of .77c. Then for values of v between .45c and .77c, $m = m_0[1 + \frac{1}{2}(v/c)^2]$; we will call this the "Mewtonian mass." Similarly, we find that for values of v between .77c and .90c we can ignore all terms beyond the third without introducing an error greater than 10%, so that, within these limits, $m = m_0[1 + \frac{1}{2}(v/c)^2 + \frac{3}{8}(v/c)^4]$; this we will call the "Pewtonian mass." In general, then, not only does relativity physics reduce to Newtonian physics for a particular range of velocities, but for other ranges of velocities it reduces to Mewtonian or Pewtonian physics and so forth.

Of course, no physicist or philosopher has yet come forth to propose a defense of Mewtonian or Pewtonian physics along these lines, undoubtedly because neither of them has been developed and widely used in the recent past. But the efforts we have examined to show that relativity theory is a generalization of Newtonian theory are based on the formal and quantitative relations between the two theories, not on their historical relation, and on formal and quantitative grounds there is no reason for maintaining that relativity theory is a generalization of Newtonian theory any more than of Mewtonian or Pewtonian theory. None of this is to be taken as denying the obvious fact that relativity theory grew out of Newtonian theory, but the point is that it grew out of the failures of Newtonian theory, not out of its successes, and the process which yielded the new physics was not a process of generalization, but rather a process of adopting new presuppositions about the structure of physical reality and transforming the basic concepts with which scientists think about and deal with the physical world.

Our discussion of the relation between classical mechanics and relativity has been, thus far, largely directed against the view that relativity is a generalization of Newtonian mechanics, and little has been said to clarify the actual relation between the two theories. Similarly, little has been said about the relation between Galilean and Aristotelian physics, although the way in which these two scientific revolutions have been juxtaposed in our discussion, along with our rejection of the thesis that Aristotelian physics was not science at all, should suggest that there is a similar relation in the two cases. But the attempt to describe more precisely what this relation is is best left for the following chapter.

Scientific Revolutions

The central notion around which this chapter has turned is that of a scientific revolution. It has no doubt struck the reader that I have been using this term in a sense somewhat different from the traditional sense. Traditionally the term has been used to refer to a single event occurring during the sixteenth and seventeenth centuries which brought modern science into existence. As the term is being used here, there have been a number of scientific revolutions, the one in the sixteenth and seventeenth centuries being only one of them, although a particularly important one since it is the first case in which a well developed and widely accepted scientific theory was overturned, and there is no reason to believe we have seen the last.

The notion of a scientific revolution is a philosophical, not a scientific, notion. It is a notion used in constructing a theory of science. But just as scientists make use of data supplied by observation and experiment in constructing their theories, so the philosopher of science makes use of the data supplied by the history of science in attempting to construct a philosophical theory of science. In Part I I argued that a philosophy of science is a theory of the same general kind as a scientific theory. We can now add a point to that argument. The change in the notion of a scientific revolution involved in our analysis of science is a change of the same kind as the changes in scientific concepts we have examined in this chapter. The new philosophy of science is an attempt to bring about a philosophical revolution and the concept of a scientific revolution, taken over from older theories of science, is changed in the process. This is perhaps most clearly indicated by the use of the term "revolutions" in the plural, a usage which makes no sense if we take "scientific revolution" to refer to a unique event. The new approach to the philosophy of science has grown out of the failure of the older approach to solve its problems and out of anomalies revealed by modern studies of the history of science. In the construction of the new approach it is clear that both the sense and reference of the term "scientific revolution" has been changed. On the other hand, at least one strand of meaning from the older notion remains: a scientific revolution is still viewed as a fundamental change in the way we think about reality.

Discovery

A central doctrine of logical empiricist philosophy of science is a distinction between the context of discovery and the context of justification. The thrust of this distinction lies in the familiar thesis that the philosopher is concerned only with logical questions and that such questions arise only after a scientific theory has been formulated; the process by which a scientist happens to think up a particular theory, it is argued, is of no concern to the logician or philosopher, even though it may be of considerable interest to the psychologist or sociologist. Rudner, for example, writes:

> Now, in general, the context of validation is the context of our concern when, regardless of how we have come to discover or entertain a scientific hypothesis or theory, we raise questions about accepting or rejecting it. To the context of discovery, on the other hand, belong such questions as how, in fact, one comes to latch onto good hypotheses, or what social, psychological, political or economic conditions will conduce to thinking up fruitful hypotheses.[1]

Similarly Reichenbach, arguing that epistemology is concerned only with the context of justification, relegates to the context of discovery any consideration of "the thinker's way of finding [a] theorem"[2] and any discussion of "actual thinking processes,"[3] and Popper maintains that:

> The initial stage, the act of conceiving or inventing a theory, seems to me neither to call for logical analysis nor to be

susceptible of it. . . . Accordingly I shall distinguish sharply
between the process of conceiving a new idea, and the
methods and results of examining it logically. As to the task
of the logic of knowledge—in contradistinction to the
psychology of knowledge—I shall proceed on the assumption that it consists solely in investigating the methods em-
ployed in those systematic tests to which every new idea must
be subjected if it is to be seriously entertained.[4]

There are two theses built into this distinction: that a sharp line can
be drawn between the discovery and the testing of scientific
theories; and that it is only with respect to the testing of theories that
we can speak of a logic at all, for logic has nothing to say about
discovery. Both of these theses are dubious. We will begin by
examining the first.

The writers quoted above are able to distinguish sharply
between discovery and testing only because they identify
scientific discovery with conceiving or entertaining a new
hypothesis. But we do not usually refer to a proposal as a discovery
unless it has passed enough tests to become, at least for a time, a
part of the accepted body of science. Kepler, for example, is
credited with having discovered the elliptical orbit of Mars, but this
discovery was the result of years of work which included the
proposal and rejection of a number of circular and ovoid orbits. Yet
in spite of the fact that he conceived or invented numerous
hypotheses, only one of them, the one which seemed finally to fit
his data within an acceptable margin of error, is considered a
scientific discovery. Similarly, Newton is credited with having
discovered the inverse square law of gravitation, and Clairaut with
having shown how this law can be used to compute the motion of
the moon's apsides. Before Clairaut solved his problem he at one
point proposed that Newton's law be modified to make the force of
gravitation inversely proportional to $1/r^2 + 1/r^4$,[5] yet we do not
credit Clairaut with having discovered this law of gravitation. Not
every idea which pops into a scientist's mind is a discovery. When
we credit Galileo or Newton or Einstein or Bohr with having made
scientific discoveries, we only consider those hypotheses which
they had good reasons for entertaining to be discoveries. The
context of justification is thus part of the context of discovery and no
sharp line can be drawn between discovery and justification. It
might be replied that even if justification is a part of discovery, we
can still distinguish two parts of the discovery process: a logical
part which we call "justification" and a creative part which is
non-logical. This takes us, however, to the second thesis referred to
above, the claim that logic has nothing to do with the creative
aspect of science.

Most of the arguments against the possibility of a logic of
discovery are directed at the traditional notion of an inductive logic
which will allow us to infer scientific laws and theories from sets of
observation statements. Against this form of logic of discovery,
Hempel argues:

Induction is sometimes conceived as a method that leads, by means of mechanically applicable rules, from observed facts to corresponding general principles. In this case, the rules of inductive inference would provide effective canons of scientific discovery; induction would be a mechanical procedure analogous to the familiar routine for the multiplication of integers, which leads, in a finite number of predetermined and mechanically formulable steps, to the corresponding product. Actually, however, no such general and mechanical induction procedure is available at present; otherwise, the much studied problem of the causation of cancer, for example, would hardly have remained unsolved to this day. Nor can the discovery of such a procedure ever be expected.[6]

While everything Hempel says about induction here is correct, the force of his argument depends on equating the notion of an inductive logic with the notion of a set of mechanical rules, and the argument becomes a general argument against the possibility of any logic of discovery only if we assume that a logic must be a set of mechanical rules. Thus Hempel goes on to argue that while deductive logic does provide us with such rules, they are not rules for the discovery of new theorems, even in a deductive science such as mathematics.[7] All that deductive logic offers is a set of standards to which any proposed proof must conform. In other words, the role of deductive logic is retrospective: it provides us with criteria for a rational reconstruction[8] which makes clear and explicit the relation of the newly proven theorem to previously accepted theorems and axioms. The only point at which mechanical rules enter into consideration at all, even in a purely deductive science, is in the process of checking whether a proposed proof is valid. Thus given the notion of logic as a set of mechanical rules, it is only in the process of checking proofs that we are engaged in a logical activity, and the thesis that there is no logic of discovery follows trivially.

Popper takes a further step. He first identifies logic with analytic propositions and tautologies, and maintains that there cannot be an inductive logic since inductive inference does not fit this model: "Now this principle of induction cannot be a purely logical truth like a tautology or an analytic statement . . . for in this case, all inductive inferences would have to be regarded as purely logical or tautological transformations, just like inferences in deductive logic."[9] He then identifies the logical with the rational and thus denies that scientific discovery is rational: "There is no such thing as a logical method of having new ideas, or a logical reconstruction of this process. My view may be expressed by saying that every discovery contains 'an irrational element', or a 'creative intuition', in Bergson's sense."[10] Thus, together with the dichotomy between discovery and justification, a further dichotomy is drawn between being rational (equals logical, equals tautological, equals mechanical) and creativity: in effect one is rational only to the extent that one wields mechanical rules; any departure from mechanical rules is a step into the realm of the irrational, or more accurately, the "arational," since the point is that we have stepped out of the realm where the notion of being rational applies, not that the canons of rationality have been violated; and, in particular, any creative act is arational.

This, I submit, is a strange concept of rationality. Rather, those decisions that can be made by the application of algorithms are paradigm cases of situations in which rationality is not required; it is exactly in those cases which require a decision or a new idea which cannot be dictated by mechanical rules that we require reason.[11] In attempting to construct a logic of discovery I will not propose a set of rules for generating new scientific theories, but rather a conceptual framework which will allow us to clarify the relation between new scientific discoveries and the existing body of scientific knowledge and problems in the context of which these discoveries were made, and will provide us with a basis for understanding the kind of reasoning which leads the creative scientist to a new discovery.

There is another direction from which our notion of a logic of discovery may be attacked. It may be argued that in the deductive reconstruction of an argument the connection between propositions is a necessary connection and no logic of discovery can meet this standard. There is no necessary connection, for example, between Galileo rolling balls down an inclined plane and his discovery of the law of free fall, and in rejecting the claim that the relations a logic of discovery seeks to analyze can be expressed in tautologies I have already conceded that a logic of discovery cannot present necessary relations. But, again, there is no reason to limit the notion of a logical relation to that of a necessary relation. Rather, I propose to take my lead from the logical empiricists' own identification of the logical with the rational and use the notion of the logic of discovery in the sense of a *rationale*, i.e., an attempt to display an intelligible structure, even though this may not consist of necessary relations.[12] In particular, I am going to develop a notion of dialectic and apply this notion to the analysis of scientific discovery.[13]

Dialectic

It must be emphasized at the outset that the notion of dialectic I will be using here is taken from Plato, not from Hegel. We will be concerned with the process by which problems arise in the context of accepted presuppositions, successive attempts are made to solve these problems, and, in some cases, changes in one or more of the presuppositions are proposed thereby altering the range of acceptable solutions. There are two central Hegelian themes that will not appear in our notion of dialectic: that each new proposal is somehow necessitated by the preceding development, and that each new proposal stands in "opposition" to preceding proposals. Before attempting a detailed analysis of specific cases of scientific discovery, it will be useful to illustrate our notion of dialectic by considering the initial attempts to define "justice" in Plato's *Republic*.

The first definition proposed by Cephalus is, "telling the truth and paying back anything we may have received."[14] Socrates criticizes this proposal by pointing out a case which fits the definition without being a case of right conduct: Suppose a friend has left a weapon with me and has since gone mad. It would not be right to return the weapon if it were asked for and the proposed definition is thus unacceptable.[15] But if Cephalus believes his proposed definition to be adequate, why should he accept Socrates' example as a counter-instance? Why not adhere to his definition and insist that it is just to return the weapon? Cephalus' decision to abandon his definition shows that the attempt to define "justice" is taking place in the context of a set of presuppositions which provide criteria for determining what constitutes an acceptable answer. The participants may not be able to state these presuppositions, but they are able to recognize instances in which they have been violated, and the thrust of Socrates' objections is to point out inconsistencies between a proposed solution and an accepted presupposition.

Once an objection has been accepted there are two ways to proceed: we can offer a new definition which will be more nearly adequate given the accepted presuppositions, or we can attempt to alter those presuppositions. Both of these alternatives are illustrated early in the *Republic*. Polemarchus' definition of justice as helping one's friends and injuring one's enemies[16] is offered as an improvement on Cephalus' definition and criticized by Socrates on the grounds that a just man will not harm anyone and that *qua* just he can do little to help his friends.[17] Thus far the controlling presupposition of the dialogue is that justice is concerned with how we treat others, and the effect of Thrasymachus' proposal that justice is the advantage of the stronger[18] is to reject this presupposition.

We have followed enough of Plato's argument to illustrate some key features of dialectic. Dialectical logic applies to attempts to answer a question, but questions, as we have already seen, only arise in the context of presuppositions.[19] In the context of a particular set of presuppositions many answers to a question are possible; there is no effective procedure for determining which answer ought to be put forth at a given point, although the detailed structure of the argument will often be highly suggestive, and many self-consistent proposals will be ruled out by the presuppositions. (No numerical comparison of the relative sizes of the permissible and impermissible sets is, in general, possible since both sets may be denumerably infinite.) Thus, in the *Republic*, not only is Polemarchus' proposal a further development of Cephalus' line of argument, but within the confines of the presuppositions accepted by them, such proposals as that justice is evading one's duties or harming one's friends or the advantage of the stronger are not acceptable.

At any point in the course of an argument, however, it is also possible to question a previously accepted presupposition and thus attempt to turn the argument in a new direction. Thrasymachus does this, but even then there is no total break in the development. Thrasymachus does not change the subject; he rejects a specific presupposition while accepting others, e.g., that an acceptable definition of "justice" must apply to all cases, and the argument continues on this modified foundation. Were he to hold no presuppositions at all in common with the others, rational debate would become impossible; it is the presuppositions the protagonists hold in common that provide the touchstone for debate.

Perhaps the most important feature of dialectical logic is that it does not deal with relations between isolated or relatively isolated propositions, but with the role of propositions and questions in so far as they are parts of structured systems of presuppositions and problems. It does not provide a set of formal rules for analyzing the relationships between statements, as does deductive logic, but let us recall that the reason why formal rules are of central importance for deductive logic is that deduction is concerned only with formal relations, not content. A dialectical logic, however, is a content logic, not a formal logic. Efforts to construct a formal logic of discovery are highly implausible (and thus all attempts at constructing a logic of discovery are implausible for those who identify logic with formal logic) exactly because it is impossible to understand the structure of scientific research without understanding its content in its historical setting. What the concept of a dialectical logic provides, then, is a tool for examining the structure of research in terms of the historical context.

Another crucial difference between dialectical logic and deductive logic can now be elucidated. A deductive logic offers only an instrument for the rational reconstruction of completed research programs. The concept of dialectic, however, can provide an instrument for analyzing both the relations between successive theories and the actual research process because it is concerned with analyzing scientific thought in terms of the intellectual tools used by the scientist. Thus we can distinguish two ways of looking at the development of science: the study of scientific discovery examines scientific research from the point of view of the practicing researcher; the study of scientific development looks back over the history of science and examines the relations between successive theories. Both these forms of analysis are necessary for an adequate understanding of science.

Scientific
Discovery

Consider first the case of the motion of the moon's apsides in Newtonian mechanics. We have seen that problems only arise if research is conducted in terms of a theory, and the problem of the moon's motion is an example of a conflict between observation and theory resulting from the attempt to account for it in Newtonian terms. At one point in his attack upon this problem Clairaut considered modifying the inverse square law to read "$1/r^2 + 1/r^4$."[20] At first glance this might seem a fundamental break with Newtonian theory but, on reflection, it is seen to be squarely within the framework of that theory. Clairaut is still seeking a mathematical expression for a force which will permit the application of the rest of Newtonian theory to the problem; logically possible alternatives such as adding an epicycle are thus excluded. In addition, the form of the proposed formula is clearly suggested by the Newtonian framework. Newton's force formula gave entirely acceptable results for the orbits of the planets; any new formula would have to give very nearly the same results. In the Newtonian context, the most obvious fact distinguishing the moon from the planets is its proximity to the earth. This makes it eminently plausible (although by no means necessary) to seek a formula which will differ from the Newtonian formula for relatively small distances from a gravitating body and "reduce" to the Newtonian formula for larger distances. Finally, Newton himself had shown how to construct formulas with the necessary properties by his use of an inverse power function, giving Clairaut good reason to consider adding another inverse power term, the significance of which would decline rapidly with increasing distance. Clairaut might have experimented with powers other than four, perhaps even with other kinds of expressions; but we can be quite sure that there is a huge number of self-consistent force functions, $F = 5/r^2$ or $F = 1/r^2 + r^2$, for example, which he would never have considered as possible alternatives to Newton's inverse square law for the simple reason that given the context in which he was working and the information he had available, it would have made no sense to do so. Clairaut eventually dropped this hypothesis and found a way of solving the problem without rejecting the inverse square law, so that his formula is not a case of a scientific discovery. But even in the case of this rejected hypothesis we can see how the scientist's research framework provides reasons for suggesting some hypotheses and not seriously considering others.

Still in the Newtonian context, consider the discovery of Neptune. Again scientists sought to use the accepted theory in an area to which it ought to apply and found a conflict between theory and observation. Again the response was to offer an hypothesis, one completely in accord with the accepted theory: that there is another planet exerting a gravitational pull which perturbs the orbit of Uranus.[21] Upon the verification of this hypothesis the discovery of Neptune was announced. The problem of the motion of Mercury's perihelion arose in the same way and the hypothesis of Vulcan's existence emerged from the same theoretical framework as did the hypothesis of Neptune's existence, so that Leverrier's proposal was a reasonable one even though Vulcan's existence was never confirmed.[22]

These examples illustrate our approach to the analysis of discovery, but do not provide a test of its adequacy. The necessary test is its ability to clarify the rationality of discoveries which lead to fundamental scientific revolutions, and it is to examples of this kind that we now turn. Our main concern is to demonstrate that there is an intelligible basis in the existing tradition for even the most revolutionary proposals, but that this in no way detracts from the originality of the creative scientist. We shall also see that while in some cases a fundamental discovery involves a wholly new hypothesis, the introduction of a new hypothesis is by no means a necessary feature of a scientific revolution.

Let us consider Copernicus' development of a heliostatic version of planetary astronomy. To begin with, we must note that he was working in a definite problem situation which he described in the preface to *De Revolutionibus*:

> First, the mathematicians are so unsure of the movements of the Sun and Moon that they cannot even explain or observe the constant length of the seasonal year. Secondly, in determining the motions of these and of the other five planets, they use neither the same principles and hypotheses nor the same demonstrations of the apparent motions and revolutions. So some use only homocentric circles, while others, eccentrics and epicycles. Yet even by these means they do not completely attain their ends. Those who have relied on homocentrics, though they have proven that some different motions can be compounded therefrom, have not thereby been able fully to establish a system which agrees with the phenomena. Those again who have devised eccentric systems, though they appear to have well-nigh established the seeming motions by calculations agreeable to their assumptions, have yet made many admissions which seem to violate the first principle of uniformity in motion. Nor have they been able thereby to discern or deduce the principal thing—namely the shape of the universe and the changeable symmetry of its parts.[23]

It was the astronomical tradition in which Copernicus worked that bequeathed him these problems, and although he broke with that tradition in a way which led to its eventual demise, he was still very much a member of that tradition and thought in its terms: besides taking over its problems, he maintained the principle that all heavenly bodies move in circles and he made extensive use of epicycles and eccentrics. He also took the center of the earth's orbit, not the sun, as the center around which the planets move.

While Copernicus developed a new approach to planetary astronomy, he did not propose an original hypothesis never thought of before. The hypothesis that the earth moves had been widely discussed and rejected throughout the history of ancient and medieval astronomy. Among the ancients, Heraclides had recognized that the apparent daily motions of the heavens from east to west could be the result of a daily motion of the earth from west to east, and Aristarchus had postulated both a daily and an annual motion of the earth.[24] Similarly, during the fourteenth century Buridan and Oresme had discussed the possibility of a diurnal rotation of the earth, the latter arguing for it in great detail before finally rejecting the hypothesis.[25] But the fact that

Copernicus did not invent a new hypothesis in no way detracts from his genius or originality. Rather, it is a mistake to tie the notion of originality to the invention of new hypotheses. The locus of Copernicus' genius is to be found in the fact that he developed the heliostatic hypothesis in detail even though it had previously been rejected and seemed prima facie implausible. We should recall Galileo's judgment that Copernicus' "sublime intellect" is celebrated because "with reason as his guide he resolutely continued to affirm what sensible experience seemed to contradict."[26]

Much the same could be said with respect to Kepler's discovery that the orbit of Mars is an ellipse, with the important exception that, as Hanson points out,[27] his decision to abandon circular orbits was one of the boldest innovations ever made, for there does not seem to be a single instance in the previous history of astronomy in which the principle of circular motions of celestial bodies had been questioned. Kepler, too, began his study of the Martian orbit with the assumption that it is a circle, and when he found that he could not fit Tycho's data to a circular orbit he first assumed that there was something wrong with his methods of computation.[28] Only very slowly and reluctantly did he give up the circle. His first new hypothesis was the ovoid, a figure which is circular at the apsides and which, like the circle, has only a single focus at which the sun could be located. That Kepler chose an ovoid, and which ovoid he first chose (for he tried a number of different ones in the course of his investigation), is neither a necessary consequence of his rejection of the circle nor an arational guess; it is a reasonable next step for Kepler to have made in his search for an alternative figure.

Due to the mathematical difficulties of the ovoid Kepler used the ellipse, a curve which is not even partly circular and which has two foci, as an approximation to simplify computations, and it was only after repeated failures with the ovoid that he adopted the ellipse as a physical hypothesis.[29] Again we find that the revolutionary thinker began with a problem generated by an existing research tradition, took his guidance from this tradition, and moved away from it only with the greatest caution and in small steps. Yet the fact that a Copernicus or a Kepler worked in this way did not prevent them from contributing to the total overthrow of that research tradition.

Turning to more recent history, let us look at the background of the theory of relativity. We saw in the previous chapter that this theory constituted a fundamental departure from classical physics; but this is not inconsistent with the claim that Einstein's new proposal was intelligible in terms of the context in which he was working and its problems. Einstein was trying to solve problems generated by classical physics, in particular classical electrodynamics. In the opening sentence of his first paper on relativity he states his starting point: "It is known that Maxwell's electrodynamics—as usually understood at the present time—when applied to moving bodies, leads to asymmetries which do not appear to be inherent in the phenomena." After giving one example of such an asymmetry, he adds:

> Examples of this sort, together with the unsuccessful attempts to discover any motion of the earth relatively to the "light medium," suggests that the phenomena of electrodynamics *as well as of mechanics* possess no properties corresponding to the idea of absolute rest. They suggest rather that, *as has already been shown to the first order of small quantities,* the same laws of electrodynamics and optics will be valid for all frames of reference for which the equations of mechanics hold good.[30]

The development here too is similar to that of the Platonic dialogue. Beginning from an existing theory, i.e., an answer to a question, Einstein gives reasons why this answer is inadequate and proposes a new solution, one which may be quite different from the one he rejects but which nonetheless takes its departure from some of the ideas of the rejected hypothesis. Just as there is no logically necessary reason why Einstein must abandon classical theory at this point rather than joining the efforts to make it work, so there is no logically necessary reason why he must seek a new approach in the particular way that he does. Still, Einstein consciously built his new theory on a foundation provided by the classical physics he was overturning. Indeed, neither of the two postulates of the special theory of relativity—that the laws of nature are the same for all Galilean frames of reference and that the speed of light is independent of the motion of the emitting source—is an entirely new principle. The principle of relativity is a generalization of accepted principles (as Einstein points out in the italicized parts of the above passage), and the independence of the speed of light from the speed of its source is a consequence of the wave theory of light and of Maxwell's equations.

But the fact that the only "new" postulate of the theory was a generalization of accepted principles does not entail that the new theory is a generalization of the older theory. As in the case of Copernicus, it is not so much Einstein's assumptions taken individually that are revolutionary as the way in which he put them together and his willingness to accept the conclusions that followed from them no matter how counter-intuitive they might seem. Indeed, from the point of view of classical mechanics the two principles on which Einstein founded his theory are mutually inconsistent, although they are consistent in the new theory because of his alteration of the traditional concept of time.[31] Perhaps the most radically new single thesis of the theory was the analysis of simultaneity and the consequent rejection of the notion that time is absolute. Yet even in this instance there was at least one other physicist, Larmor, who preceded Einstein in proposing the formula for the time dilation of a moving body.[32]

Finally, Einstein was not alone in being aware of the problems of classical physics, nor in considering the possibility of a fundamentally new approach. Poincaré perhaps came closest to preempting him. In a paper read in 1897 in Zurich (where Einstein was a student, although it is not known that he attended the conference), Poincaré argued that "Absolute space, absolute time, even Euclidean geometry, are not conditions to be imposed on mechanics; one can express the facts connecting them in terms of non-Euclidean space."[33] And in 1904 he suggested: "Perhaps we should construct a whole new mechanics, of which we only succeed in catching a glimpse, where, inertia increasing with the velocity, the velocity of light would become an impassible limit."[34] That others were aware of the possibility and perhaps even the desirability of a new departure further demonstrates that more was involved in Einstein's innovation than the arational proposal of a new hypothesis; his greatness lies in the fact that he did not merely offer a new hypothesis or suggest a new approach, but rather constructed a new physics. The unfortunate tendency to attribute a discovery to the thinker who first ennunciated a new hypothesis, rather than to the one who put the pieces together, apparently lies at the source of Whittaker's notorious attribution of the development of the theory of relativity to Lorentz and Poincaré while giving Einstein credit only for the new formulas for aberration and the Doppler effect.[35] The same can be said of Newton's famous battle with Hooke over who first stated the inverse square law; even if Hooke first formulated the hypothesis, it was Newton who developed a new physics.

Scientific Change

Our notion of dialectic will enable us to resolve a central problem introduced in the last chapter, the nature of the relationship between successive scientific theories. As we have seen, no one has seriously maintained that new theories are deduced from old ones, and the view that new theories are generalizations of old theories will not stand up. The claim that older successful theories can be deduced from later theories as limiting cases (a corollary of the generalization thesis) fails as well. Revolutionary theories involve different fundamental presuppositions and concepts and a different way of looking at reality than the theories they supersede. But if this is the case, how can there be any logical relation between successive theories and, more importantly, on what grounds can successive theories be compared?

The key to these questions lies in our notion of dialectical change. A dialectical change from one theory to another is a reorganization of the strands of the theoretical web, along with the removal of some strands and the addition of others; this reorganization accounts for the changes in the meaning of scientific concepts and observations associated with a scientific revolution, since both the concepts and the data of a theory derive their meaning from their location in the theoretical web. But theory change takes place within definite problem situations, within which the strands retained in the new theory provide the continuity of development and the grounds for comparison even though these strands take on different meaning in the new theoretical structure.

Much discussion of theory change has been confused by a failure to distinguish between two different theses: the thesis that if we are to have a rational basis for choosing between rival theories at a particular moment in the history of science, we must have some standard to appeal to which is accepted by proponents of both theories, and the thesis that if scientific change is to be rational there must be some eternal standard against which we can compare any theories. Proponents of the latter view have sought their eternal standard either in some set of general principles independent of any specific theory or in a set of theory-independent observations or laws; sometimes a writer appeals to both kinds of criteria at the same time. Among the supra-scientific principles that have been used in traditional philosophy of science are such supposed a priori principles as the principle of causality, of sufficient reason, or of the uniformity and simplicity of nature. More recently, we find a number of philosophers trying to construct an inductive logic as well as Popper's attempt to dictate a set of methodological rules to which science must conform. The main thrust of these suggested criteria is that they have nothing to do with any particular theory, and therefore can serve as criteria by which theories can be evaluated. Similarly, empiricists have maintained that the ability to account for the widest range of possible observations can serve as a standard because these observations are taken to be free of the influence of any theory. It is in this vein that Feigl, responding to claims that scientific observations are theory-laden because of the instruments used in making them, suggests that "it is the domain of the elementary, rather directly testable empirical laws (instead of the 'given,' be it conceived as sense data or as perceptual Gestalten) that is the testing ground to which we should refer in the rational reconstruction of the confirming or disconfirming evidence for scientific theories."[36]

Yet, as we have seen, the appeal to observations and to empirical laws cannot be made independently of theory, even principles such as universal causation are integral parts of scientific theories which can be questioned as a result of later scientific developments, and while methodological rules such as an inductive logic are indeed independent of any scientific theory, they are integral parts of particular philosophical interpretations of science and thus no more capable of providing eternal standards than a specific scientific postulate. The choice between scientific theories does not take place by appeal to eternal standards, established by philosophers, but rather by appeal to scientific standards which are provided by the theories involved. In order to develop this point we must return, again, to the examination of specific cases of scientific change in their historical circumstances.

Let us reconsider Einstein's introduction of the theory of

relativity. Since most physics students study relativity after Newtonian theory, it is common for textbooks to develop relativity theory in terms of the kinematics and dynamics of particles and only later introduce its significance for electrodynamics, often almost as an afterthought. The physics of particles and the cosmology which, from a Newtonian viewpoint, is constructed on the basis of particle dynamics are the fields in which the radical differences between relativity theory and classical physics are most dramatically apparent and around which recent philosophical debate has centered. But it is often forgotten that the initial impetus for Einstein's development of his theory lay in problems raised by electrodynamics: the title of the first relativity paper is "On the Electrodynamics of Moving Bodies," and it begins "It is known that Maxwell's electrodynamics—as usually understood at the present time—when applied to moving bodies, leads to asymmetries which do not appear to be inherent in the phenomena."[37]

The paper is divided into an introduction, in which Einstein formulates the two principles of the theory, and two main parts. Part one deals with kinematics: here Einstein develops his analysis of simultaneity and deduces the relativity of length and time, the Lorentz transforms and the rule for the composition of velocities. But part I is only a preliminary to the electrodynamical part; nowhere in part I is the new kinematics applied to material particles. This application is made only towards the end of part II and only because particles in general can be taken to be electrons. Thus having deduced the formula for the effect of velocity on mass for charged particles, Einstein adds: "We remark that these results as to the mass are also valid for ponderable material points, because a ponderable material point can be made into an electron (in our sense of the word) by the addition of an electric charge, *no matter how small*."[38] Similarly, after deducing the expression for the kinetic energy of a moving electron, he writes: "This expression for the kinetic energy must also, by virtue of the argument stated above, apply to ponderable masses as well."[39] But Maxwell's equations for electrodynamics were not replaced by new equations as were the equations of Newtonian mechanics. Rather, "asymmetries" in electrodynamic theory were removed, the troublesome ether was eliminated, and Maxwell's equations were shown to be invariant with respect to Galilean reference frames.

To a large extent, then, Einstein's project can be viewed as an attempt to save electrodynamics. Although Maxwell's equations take on a new meaning in the context of the new theory (they no longer describe waves propagated in the ether, for example), the Maxwell equations themselves provide one of the standards, acceptable to both classical and relativistic physics, on the basis of which they can be compared. This is not an eternal or theory-free standard, e.g., one which could have been appealed to in the dispute between Aristotelian and Galilean mechanics. Rather, it is a standard internal to physics which serves as a point of mediation in this dispute.

Without attempting to be exhaustive, let us look at one more crucial standard which both theories accepted: the demand for precise numerical predictions. One key test that relativity had to pass was its ability to provide numerical predictions that are essentially the same as those of Newtonian physics for slowly moving bodies and more accurate for bodies of velocity approaching that of light. We have seen[40] that the ability of two theories to predict the same values for quantitative parameters is consistent with their having different conceptual content; two related points may now be made. The first is that because both theories accepted the standard of accurate quantitative predictions as well as many of the customary observational procedures, it was possible to use quantitative results obtained from observation to test the two theories even though these results had different meanings in the context of each. The second is that the standard of quantitative accuracy is by no means a universal feature of the history of science. Galileo, for example, could not invoke it without extended prior argument, since the very relevance of mathematics to physical problems was one of the points he was disputing with the Aristotelians.[41] The only area of research which was already fully quantitative in Galileo's day was mathematical astronomy, but this was viewed solely as a system of computation devices, not as a science. Physical astronomy, i.e., the Aristotelian picture of celestial bodies supported by transparent spheres, was generally recognized to be inconsistent with the epicycles used in computing the positions of the planets, and it is the non-quantitative Aristotelian theory that was taken to be the scientific theory. Similarly, the medieval attempts to explain projectile motion were purely qualitative and here too Galileo fought to introduce quantitative considerations.[42] Again we find that the contemporary state of physics provides criteria for deciding between two competing fundamental theories, although these are not universal criteria accepted independently of the history of science.

For another example let us consider further the debate between Galileo and the Aristotelians. Again two radically different theories providing quite different pictures of the nature of the physical world are under debate. When we examine the dispute we find that at least some of the mutually acceptable phenomena to which the protagonists on both sides appealed seem to be of the order of Feigl's "elementary, rather directly testable empirical laws." To take one of the most important cases, Galileo and his Aristotelian opponents recognized that a stone dropped from a high tower falls "straight," i.e., parallel to the tower, and agreed that an adequate physics must account for this "law." But we cannot view such laws as permanent bits of knowledge which are independent of any physical theory. The observed phenomena in question had a completely different significance in the context of Aristotelian and Galilean dynamics. For an Aristotelian the "straight fall" of a stone was an instance of a body returning to its "natural place," defined as a particular portion of space, and the real motion of the body (a concept of fundamental importance to both Galileo and the Aristotelians, although meaningless in the context of relativity theory) is accelerated motion in a straight line. For Galileo, the "straight fall" was only one of the motions of the body, the second being the circular inertial motion of the earth in which the stone and tower share; the real motion of the stone is compounded of these two motions and is neither straight nor accelerated. Rather, it is uniform circular motion to the stone's "natural place," where this notion is now defined by reference to a particular material body, the earth. Further, the motion is along a semicircle which has the top of the tower and the center of the earth at the ends of its diameter.[43] Thus there is a class of observed events which both theories, each for its own reasons, recognizes must be explained; and these mutually acceptable observations serve as points of comparison between them even though they are instances of very different laws in the context of the different theories.

Nor can we retreat back to observations, independently of the laws they instantiate, to find some permanent basis against which all theories can be tested, for the theories determine what kinds of observations are relevant, and this too changes as theories change. For example, one of the standard arguments against the diurnal rotation of the earth, an argument used by Tycho Brahe, among others, was that if the earth rotates from west to east, there ought to be a continuous wind from east to west. The observed absence of such a wind was taken as a counter-instance to the purported rotation. Galileo accepted this as relevant and attempted to turn it into an argument in favor of the motion of the earth. He argued that the air does not participate in the circular motion of the earth because it is not an earthy body. Galileo seems to accept here a form of the Aristotelian theory of natural kinds which have different physical properties so that neither the natural circular motion nor the circular inertial motion which Galileo attributes to the earth can be invoked to explain why the air moves along with the earth. Rather, Galileo argues that the air is carried along with the earth only because of the roughness of the earth's surface and because it is mixed with "earthy vapors"[44] which share the properties of the earth. Where there is no roughness and a smaller quantity of earthy vapors, e.g., over large bodies of water, Galileo argues, we should find a continual breeze from east to west that is strongest at the equator where the rotation is fastest. This conclusion from the Copernican theory, he continues, is confirmed by experience:

> For within the Torrid Zone (that is, between the tropics), in the open seas, at those parts of them remote from land, just where earthy vapors are absent, a perpetual breeze is felt moving from the east with so constant a tenor that, thanks to this, ships prosper in their voyages to the West Indies. Similarly, departing from the Mexican coast, they plow the waves of the Pacific Ocean with the same ease toward the East Indies, which are east to us but west to them.[45]

Similarly, Galileo considered his theory of the tides to be his strongest argument for the motion of the earth. Water, too, is a distinct element which does not participate in the natural motion of the earth, but is rather, "because of its fluidity, free and separate and a law unto itself."[46] The tides are the result of the independent movement of the water relative to the combined daily and annual motions of the earth. In Galileo's problem situation his use of ocean winds as evidence for the motion of the earth was designed to turn the tables on his opponents, and his explanation of the tides seemed to him a powerful new argument. (Originally Galileo intended to title the book *Dialogue on the Tides* but was forced by the censors to abandon that title.) But while both of these classes of phenomena, the tides and the ocean winds, still exist, to the post-Newtonian they are irrelevant to the question of the motion of the earth. Thus we find again that, in a dispute between two radically different physical theories, there are a number of specific phenomena that the proponents of both theories accept as relevant and which can serve as the basis for rational debate. But which phenomena are accepted and what criteria are applied in evaluating the theories are determined in the specific problem situation, not by universally valid a priori considerations.[47]

Looking back at these two examples, we can see that not only is the process of discovery a dialectical one, but that the relation between an older theory and the one which supersedes it is also dialectical. The new theory is continuous with the older one, for it grows out of the failures of the older theory to solve its own problems and to account for the entire range of phenomena that the latter itself selects as relevant, and it takes over many of the observations, techniques and principles of the older theory while it changes their meaning. Nothing could be more misleading than to attempt to elucidate the nature of scientific change by comparing Aristotelian physics with fully developed classical mechanics or even with Newton's *Principia*, for there were many stages and many thinkers leading from the physics of, say, 1600, to Newton and there is precious little that even Newton and Galileo hold in common. When we examine a dispute such as that between Galileo and the Aristotelians in its own historical context, we find that Galileo introduced such radical innovations as the concept of inertia and the law that the velocity of a falling body is proportional to the time of fall and not the distance. But we must remember that he also retained very Aristotelian notions of natural place and of the elements and that while he attacked the traditional distinction between the physics of the heavens and the physics of the earth, he did so by extending the natural circular motion of the heavens to the earth and was so far from doubting that the motions of celestial bodies are circular that he simply ignored Kepler's ellipses.

Nevertheless, while retaining many theses from traditional physics, he assembled a new structure which included both new theses and a new organization of older views. In short, Galileo's physics was both continuous with Aristotelian physics and radically different from it.

Chapter Ten

Toward A New Epistemology

It has been argued that both science and philosophy of science consist of research projects structured by some set of presuppositions: presuppositions about the nature of an aspect of reality in the case of scientific research, and presuppositions about the nature of knowledge in the case of philosophy of science. In Part I we examined the theory of knowledge upon which logical empiricism is based. It is now time to formulate the main features of the alternative theory of knowledge implicit in the new philosophy of science.

Rationality

Central to most traditional work on the theory of knowledge has been a distinction between knowledge and belief, it being assumed that beliefs can be either true or false while knowledge can only be true. If I claim to know that some proposition is true, for example, and further evidence shows that the proposition is false, we would not conclude that I had false knowledge, but that I did not know at all. Thus knowing is, by definition, infallible. A large portion of the history of philosophy consists of attempts to show how knowledge can be attained. It is because infallibility is presumed to be a defining characteristic of knowledge that Plato, in the *Theatetus,* can raise the question, "What is knowledge?" without ever asking if knowledge is infallible, but rather use infallibility as one of the criteria by which he evaluates proposed answers to his question.[1] The central role in philosophy of the quest for infallibility is equally well illustrated by the persistent search for some indubitable foundation on which to build the edifice of knowledge, and by the ease with which a writer like Hume can generate a sceptical position merely by pointing out that no "matter of fact" proposition is known to be necessarily true since its negation is not self-contradictory. Thus the doctrine that knowledge must be infallible has been a fundamental presupposition for philosophers in the same sense in which the central principles of Ptolemaic or Newtonian theory, for example, have been fundamental presuppositions in astronomy. This principle has set the primary problem of epistemology: the search for indubitable knowledge; and has provided a criterion for an adequate solution of that problem: we have achieved knowledge only when we have a set of indubitable propositions.

In general the quest for infallibility can be divided into two subproblems: the search for an indubitable starting point, and the search for infallible means of reasoning from a set of premises. These have been central to logical empiricism. We have examined the logical empiricists' attempt to take perceived data as the foundation of knowledge in our discussions of the problem of theoretical terms[2] and the relation between perception and theory.[3] Now we must consider the status of infallible reasoning procedures.

There is an important respect in which one of the main goals of traditional philosophy of science has been to remove the scientist from the decision making process and replace him with a set of algorithms. As in any other field, the aim is to approach infallibility by eliminating human judgment. For human judgment is notoriously fallible, while some of the most important accomplishments in the history of thought have been the discovery of algorithms. Only because we do arithmetic, for example, by the application of strict, unambiguous rules can we be confident of our results; if we had to rely on the judgment of some human being, rather than on a set of rules, for the answers to long division problems, long division would be a very uncertain process. This ideal controlled early logical positivist ideas on the verification of theories, receiving its most extreme expression in Wittgenstein's attempt to reduce all propositions to truth functions of atomic propositions. If Wittgenstein's goal had been attainable, the question of whether a scientific theory is true could be decided much as we determine the sum of a column of numbers. This program has, we have seen, been abandoned and replaced, among logical empiricists, by the search for an inductive logic based on probability theory. Again the project is to find an algorithm on the basis of which we can evaluate scientific theories, the assumption being that even if we cannot prove the final truth of an hypothesis, we can produce a set of rules which will allow us to determine the degree to which it has been confirmed by the available evidence.

The same ideal controls the falsificationist view of scientific procedure. Popper, realizing that no finite procedure can prove a scientific theory true, noted that the logical principle *modus tollens* provides an algorithm which, given appropriate basic statements, could prove a theory false. As Kuhn points out, Popper has, "Despite explicit disclaimers, consistently sought evaluation procedures which can be applied to theories with the apodictic assurance characteristic of the techniques by which one identifies mistakes in arithmetic, logic, or measurement."[4] Indeed, we have seen that Popper considers the entire realm of scientific discovery, which he recognizes cannot be reduced to a set of algorithms, irrational.[5]

Now the thrust of the historical cases we have examined is that there is no clear, simple relation between the results of experiment or observation and scientific theories. Even in the simplest, most straighforward instance, the case of an observational result which contradicts a theory, the practicing scientist is not bound automatically to reject part of his theory. None of the observations, for example, which seemed to show that planets do not move in circular paths around the earth were themselves sufficient to refute the principle that all celestial motions are circular. Rather, they were the source of research problems, phenomena to be explained by the theory. Similarly, the absence of stellar parallax, which was taken as a clear counter-instance to the moving earth hypothesis by contemporaries of Aristarchus and of Copernicus, and even by such an important figure in the development of modern astronomy as Tycho Brahe, was taken by Copernicus and Galileo as evidence that the distance to the stars is much greater than had been assumed, and as a research problem by later astronomers. In the same way, Leverrier and Adams' use of the apparent counter-instance of the orbit of Uranus to predict the existence of Neptune was a major triumph for Newtonian theory, while the perturbations of the orbit of Mercury became a major counter-instance contributing to the eventual downfall of Newtonian theory. The decision as to how a discrepancy between theory and observation is to be handled requires a judgment by scientists. The decision cannot be made for them by the simple application of an algorithm, and, as the history of science adequately shows, the decision procedure is fallible.

This, I take it, is the thrust of some of Kuhn's most widely attacked claims, e.g., that such questions "can never be settled by logic and experiment alone,"[6] and "the competition between paradigms is not the sort of battle that can be resolved by proofs."[7] Such statements have led many philosophers to charge Kuhn with irrationalism, a charge which Kuhn rejects, responding that it is his opponents' concept of rationality that is awry. Faced with a disagreement of this sort, we should at once suspect that we are dealing with a quarrel resulting from different sets of presuppositions, in this case involving different concepts of rationality. We must, then, examine these rival concepts of rationality.

The attempt by logical empiricists to identify rationality with algorithmic computability is somewhat strange, since it deems rational only those human acts which could in principle be carried out without the presence of a human being. But there are no grounds for such an identification, since it is possible to act irrationally while following an algorithm. Given any set of premises, it is possible by the rote application of deductive logic to deduce an infinite number of conclusions. Indeed, this could be accomplished solely by the continued addition of disjuncts. Yet a scientist who attempted to deduce testable consequences from a hypothesis in this way would be acting irrationally, even if he violated no law of deductive logic and every conclusion he arrived at followed necessarily from his premises. In what does the irrationality of this approach lie? It lies in the failure to take into account available information which cannot be applied to the problem by any known algorithm, but which provides good reason for believing that the addition of disjuncts is not a fruitful way of developing hypotheses. There are many different directions in which the scientist can proceed in attempting to deduce testable consequences from his hypothesis, each of which may be strictly in accordance with a set of algorithms, but he has no algorithm for determining which line to pursue. An informed judgment is required and it is in making such judgments that we must rely on reason. As long as decisions can be carried out by means of algorithms, human intervention is not necessary; it is exactly when we have no effective procedures to guide us that we must turn to an informed, rational human judgment. This in turn suggests another reason why the development of algorithms is important: when we establish one for dealing with a problem it is no longer necessary to devote human thought to that problem and our efforts and reason are freed to work in other directions. It is the case in which we must rely on human judgment that I propose to take as the paradigm of a situation in which reason is required.

This suggestion will seem odd to many contemporary philosophers, but it is not a new one. It is central to the notion of the man of practical wisdom in Aristotle's ethics, and a comparison with that idea will help clarify the model of rationality I am proposing.

For Aristotle, ethics, strictly speaking, is not a science. Science is the deductive demonstration of necessary truths from premises which are themselves both necessarily true and known to be true. But ethics is concerned with human behavior and, because of the complexity of human behavior, there are no first principles on the basis of which to construct a science. Ethical decisions require deliberation, the ability to weigh information and make decisions in cases in which there is no necessary knowledge. We do not, Aristotle points out, deliberate about "the incommensurability of the diagonal and the side of the square"[8] since we have a demonstration that this is necessarily the case. Similarly, we do not deliberate about how to solve a long division problem, we simply apply the appropriate algorithm. But when we lack necessary knowledge, as in the case of human behavior, an intelligent decision on how to act requires deliberation by someone who has sufficient experience of human action to deliberate well. The conclusion is not infallible and there is no guarantee that all adequately informed people who deliberate on an issue will reach the same decision, but this does not make the decision arbitrary or irrational, and the fact that equally qualified people may disagree does not imply that everyone is qualified to hold an opinion. Only those who have achieved practical wisdom, who have had sufficient experience to understand human behavior and have developed their ability to deliberate, are qualified to make ethical decisions.[9] My proposal, then, is to take the man of practical wisdom as a model of the maker of crucial scientific decisions which cannot be made by appeal to an algorithm, and I offer the making of these decisions as a model of rational thought. It is the trained scientist who must make these decisions, and it is the scientists, not the rules they wield, that provide the locus of scientific rationality.

Another concept from Aristotle's ethics, that of equity, will help to clarify further our model of rationality. Aristotle gives the following characterization of equity: "A correction of the law where it is defective owing to its universality."[10] The problem Aristotle is concerned with is that we sometimes encounter a situation that falls under existing laws, so that justice requires that we act in accordance with the law, but in which it seems unfair to apply the letter of the law. This may occur because in the formulation of universal laws it is impossible to foresee and make provision for every circumstance. The man of practical wisdom must be able to recognize this and correct the universal law in accordance with the demands of a particular situation.

An analogous point holds in the scientific case. Suppose, for example, we adopt a methodology which requires us to reject any theory inconsistent with a well confirmed falsifying hypothesis, and regard the postulation of Uranus or of the neutrino in this light. We can now view Leverrier and Adams or Pauli and Fermi as scientists who applied the general rule to the particular case and judged that although the rule applied, the case in question required special consideration and thus the rule was suspended.

It is the ability to decide how an exceptional case should be handled that is characteristic of rationality. That logic and experiment alone cannot decide the fate of theories does not imply that these decisions are irrational. It implies that they require judgments in which the results of logic and experiment are taken into account along with all that the scientist knows about the current state of his discipline. The results of logic and experiment must themselves be evaluated. It is the task of the skilled scientist to carry out this evaluation, and such evaluations furnish paradigm cases of rationality.

Similarly, rather than declare the process of scientific discovery to be irrational, we should consider the scientist actively seeking the solution of a problem as providing another paradigm case of rational thought. A mathematician, for example, who is attempting to prove a proposed theorem is, by our model, engaged in a rational activity while a mathematician who is merely checking whether a proposed proof has been validly constructed has little need for reason. The notion that there is no rational basis for discovery is plausible only if we identify the discovery of a new hypothesis with its appearance in the scientist's mind *ex nihilo,* but we have seen that this is not correct. Newton, Einstein, Bohr, Schrödinger were all striving to solve definite problems within a definite intellectual context. Even so-called accidental discoveries such as Roentgen's discovery of X-rays or Fleming's discovery of penicillin lend support to this thesis. To be sure, there is nothing rational about the accidental appearance of a mold in a culture medium of the spoiling of a photographic plate, but it requires rational thought of the highest kind to recognize that the event in question may be significant and pursue its implications.

While Aristotle's man of practical wisdom offers a model of individual rationality, scientific decision making is more complex. A central characteristic of our new model of rationality is that it recognizes that different thinkers can analyze the same problem situation and come to contrary conclusions without any of them being irrational. But the fact that a theory is arrived at rationally is not sufficient to make it a part of the body of science; that requires not an individual but a group decision. No thesis becomes a part of the body of scientific knowledge unless it has been put before and accepted by the community of scientists who make up the relevant discipline. Kuhn has provided a particularly clear description of this process:

> Take a *group* of the ablest available people with the most appropriate motivation; train them in some science and in the specialties relevant to the choice at hand; imbue them with the value system, the ideology, current in their discipline (and to a great extent in other scientific fields as well); and, finally, *let them make the choice.*[11]

It is the consensus of the workers in a discipline that determines what constitutes knowledge in this discipline, but the group may later discover that it made a mistake. The group is no more infallible than the individual (but this does not mean that it is as fallible as the individual).

The model of the man of practical wisdom applies to group as well as to individual decisions: these are judgments made on the basis of information and experience, but without the benefit of necessary truths or algorithmic procedures which can guarantee a decision immune from being overturned by a future research. Scientists can duplicate experiments and check computations and interpretations and thereby largely eliminate the charlatan and the incompetent, as well as simple and subtle errors. This does not guarantee that the community will not make mistakes; it often has. But the scientific community has been notably self-correcting. The very fact that we can list examples of group error, such as the acceptance of earth centered astronomy and the phlogiston theory of combustion, or the dismissal of mesmerism and continental drift, is evidence of this capacity for self-correction. It could be argued, in good Cartesian fashion, that, given the instances in which science has been mistaken, we can never be certain that science is correct. But Descartes' goal, the construction of a system of indubitable knowledge, has been rejected here, and only if we were to accept that goal would the Cartesian argument have any force. If we use the term "truth" in the traditional sense, reflection on the history of science, especially its recent history, gives good reason to expect the latest scientific theories to turn out false, i.e., to be rejected in the future. This suggests that we must reconsider the concept of truth and the related concept of knowledge, at least as they apply to science.

Scientific Knowledge and Scientific Truth

Knowledge and truth are epistemological concepts, concepts which occur in a theory of knowledge. If, as we have argued, there is a close parallel between the structure of scientific theories and that of philosophical theories, in the process of rejecting an older epistemology and constructing a new one we should expect some of the key concepts to change much as they do in the case of scientific theories.[12] Beginning with knowledge, we will retain a central strand of the traditional concept and take it to be the highest cognitive state, but we must reconsider what counts as scientific knowledge in the light of our analysis of science.

Without grounds for maintaining that any scientific claims are infallible, we must either deny that there is such a thing as scientific knowledge or free that concept from the concept of infallibility. Now scientific knowledge at any period consists of a number of elements: the fundamental theories which guide research, and with them the body of laws, fundamental constants, and observations that are of particular significance in the light of the guiding theory. But for a claim to become a part of scientific knowledge it is not enough that it pass formal tests, for it is common for a claim to pass the accepted tests and still be ignored because it is not judged to be of any significance. It then will not *function* as scientific knowledge. A proposed universal constant, for example, will not be used in computations, no one will make the observations suggested by a proposed theory, and so forth.[13] Thus scientific knowledge in any era is what the scientists actively take as such, and the scientific knowledge of one era may be rejected as error in the next. But the rejection of previously accepted claims will itself be made on the basis of the currently accepted views, which are themselves fallible.

This analysis may be attacked as relativism and historicism, but, conceding for the moment that these characterizations are correct, it is difficult to know how to avoid them and still make sense of science. Traditional anti-relativism amounts to a claim that only propositions which are true, in the ultimate sense, can be a part of science; but the problem remains of how we are to establish which of the propositions that make up the accepted body of science are indeed true and which are false. This is, of course, the task of scientific research, and we are back where we started. Unless scientists have an effective method for determining once and for all which propositions are true, we cannot determine which part of currently accepted science is indeed knowledge, nor even whether there exists any scientific knowledge at all. We find ourselves again in the dilemma we have seen Nagel in[14] when he maintained that the premises of scientific explanations must be true but need not be known to be true. This left us not knowing whether there are or ever have been any scientific explanations. Similarly, if we hold that only claims which have been established in such a way as to make future refutation impossible are scientific knowledge, then at best we may have scientific knowledge or not, without knowing whether we do. At this point the very concept of scientific knowledge becomes worthless.

There would seem to be three possible ways to meet this problem. One would be to scrap the concept of scientific knowledge and its related concepts and find some new way of thinking about science. This would require the development of a wholly new epistemology, something which may well be intellectually impossible. The second possibility is to continue to do philosophy of science in terms of the logical empiricist research project. If, for example, we could construct an inductive logic, we would at least be able to judge how close we have come to the achievement of knowledge. But we have seen that this research project has been carried on unsuccessfully for a long time, and that the foundations of logical empiricism are very shaky. Still, there is no proof that the logical empiricist research project must be abandoned, and as philosophers of science we are in precisely the same situation as the scientist who must decide, without benefit of an effective procedure, whether to continue to try to make the older project work or seek a new one.

The third possibility, adopted here, is to swallow the charges of relativism and historicism if need be and make them the basis for a new philosophical research project to be built on the failures of logical empiricism, a project which begins by accepting the analysis of scientific knowledge as the fallible body of accepted science. On reflection, however, there is not very much to be swallowed, since the charge of relativism entirely loses its force if one does not accept an absolutist epistemology. Only in contrast with the concept of knowledge as a system of finally established truths will the proposal to take fallible, even probably erroneous, claims as knowledge seem weak. Once we free ourselves from the belief that science can establish final truths, and accept instead that the best that science can hope to attain is tentative rational consensus on the basis of available evidence, to call this relativism in a derogatory sense becomes pointless. The same response applies to the charge of historicism. If it is historicism to maintain that what the scientific community accepts as scientific knowledge is scientific knowledge, then it is difficult to know what the objection is unless it is asserted that there are other standards to which we can appeal; then, of course, those standards must be formulated and justified. In the absence of criteria which allow us to distinguish infallibly between claims which are or ought to be accepted once and for all and those which ought to be rejected as false, we have no intelligible choice other than to take the fallible body of accepted science to be scientific knowledge.

The situation is more complex with the concept of truth. There is a clear sense of the term, what I have referred to as the "absolute" or "ultimate" sense, in which to say that a theory is true is to assert that it gives a correct description of an aspect of reality. This notion of truth, which we will refer to as "$truth_1$," denotes the goal for which scientists strive in constructing theories, but it has no relevance for the evaluation of theories since theories provide the only access we have to reality. To discuss meaningfully the truth or falsity of actual scientific theories, we need another sense of "truth" which will be referred to as "$truth_2$." In introducing this term I propose to maintain the traditional tie between the concepts of knowledge and of truth, so that whatever is known must, by definition be $true_2$, but to change the direction of the definition. Traditionally the concept of truth has been primary and "knowledge" has been defined in terms of "truth." I propose to take the concept of scientific knowledge, as it has been analyzed here, to be fundamental and introduce "$truth_2$" in terms of "knowledge": any proposition which is a part of the body of scientific knowledge is a $true_2$ proposition. Analogously, we need two concepts of falsity: a proposition is $false_1$ if and only if it does not provide an adequate description of reality; it is $false_2$ when it is rejected by the current consensus.

We can now hold quite unproblematically that scientific knowledge is both $true_2$ and known to be $true_2$ provided it is clearly understood that a theory which is $true_2$ at one time can be $false_2$ at another, and conversely. The theory of continental drift, for example, was overwhelmingly rejected by geologists in the 1920's and was thus $false_2$; during the past ten years it has been incorporated into the widely accepted theory of plate tectonics and is now $true_2$; whether it is $true_1$ remains an open question.

When a theory is the subject of current dispute, one of two situations may obtain. Scientists may consider a hypothesis but not have adequate grounds for accepting or rejecting it. Here no commitments have been made and the hypothesis is neither $true_2$ nor $false_2$. This seems to be the situation with respect to attempts to reduce all sub-atomic particles to quarks. Alternatively, there may be sizeable groups within a discipline who accept different theories. This was the case, for example, with geocentric and heliocentric astronomy for a large part of the seventeenth century, and with Newtonian and Cartesian physics in the early eighteenth century. Here we must say that one of the theories is $true_2$ for one group and the alternative theory is $true_2$ for the competing group. At most one of the competing theories can, of course, be $true_1$.

Even with the subscripts, the claim that a true theory was once, in some sense, false, or that a false theory was in some sense true, will seem paradoxical to many readers. The reason for this, however, is that what I have labelled "truth$_1$" is the only sense of truth in traditional epistemology (as well as in ordinary language). But our examination of scientific knowledge has shown that the traditional body of epistemological concepts is not adequate for the analysis of the ways in which scientific theories develop and are accepted and overthrown. It is in order to remove this inadequacy in the traditional conceptual machinery that we have introduced a new sense of "truth." It should be noted, however, that we have not introduced a wholly new concept *ex nihilo*, but built it up by modifying an older concept, one which is still retained in our conceptual web. Since we are concerned here only with epistemological questions, I will henceforth use "true" and "false" without a subscript wherever "true$_2$" or "false$_2$" is intended.[15]

Objectivity

It will undoubtedly be objected that our analysis of scientific knowledge constitutes an attack on the objectivity of science. For, it will be argued, in holding that the ultimate decision on scientific questions rests with the scientific community, rather than with an impersonal testing procedure, we introduce subjective factors into the confirmation process. Further, and more deeply, the very notion of objectivity is inconsistent with the idea of a body of knowledge based on fallible presuppositions since it leaves scientific knowledge without a foundation. Simply by altering presuppositions we change not only the body of scientific knowledge, but the kinds of questions scientists ask and the standards for judging what is scientific. Thus science becomes an arbitrary construct and there is no reason to take any proposed body of theory to be more valid than another.

The main difficulty with this objection is that it is based on the very epistemology that we have rejected. It assumes the Baconian thesis that judgments made on the basis of presuppositions are dubious, but the objection is pointless once it is acknowledged that all judgments require presuppositions. The thesis that science is objective in the described sense is not self-evident, nor is it a claim for which empirical evidence has been provided. Rather, it is a paradigmatic proposition, a basic assumption of the logical empiricist research program. To reject this program is not to reject the view that science is objective, but to pose the task of providing an alternative analysis of scientific objectivity. As in other cases in which it was necessary to redefine a concept, our new concept will be built on strands taken over from the older version.

One such strand is the dichotomy between the objective and the arbitrary, but we must reject the notion that a theory is arbitrary unless it is based on an indubitable foundation. To accept a theory because it solves some problems, eliminates others, and provides a guide to further research, is not to decide arbitrarily to accept that theory. Similarly, a theory which was once widely accepted may be overthrown because it has failed to solve its own problems, because it no longer provides a clear guide for research, and because an alternative theory has been developed which is judged to meet these criteria. But to judge that a theory is to be rejected on these grounds is not to make an arbitrary decision.

A second strand which our new concept of objectivity shares with the traditional one is that objective theories must be intersubjectively testable. This requirement is even more basic to our approach than to traditional philosophies of science. If there did exist a "scientific method" which scientists wield, the only purpose of intersubjective testing would be to assure that human frailty does not interfere with its operation; intersubjective testability would not be a logically necessary part of research. But for our approach, which requires that proposals be evaluated and accepted by the community of qualified scientists before they can become a part of science, intersubjective testing is crucial.

Finally, we must remember that scientists are trying to understand a reality which is objective in the sense that it exists independently of their theories. Scientific theories are tested against this reality which plays a central role in determining what we observe, and which is continually throwing up anomalies to remind us that our current theories are not perfect and may have to be abandoned in favor of some new approach. We have seen that we can attempt to understand physical (or biological or mental) reality only in terms of some set of presuppositions, but unless the structure of our presuppositions meshes to some degree with the inherent structure of reality, it will fail as a guide to research and be rapidly eliminated.

Descriptions and Norms

There is another, particularly complex, problem raised by the thesis that scientific knowledge is the consensus of the scientific community. There are many ways in which a consensus can be established and it is not clear that all of these are legitimate, nor is it clear that all of the factors which may play a role in bringing about consensus on a particular issue are relevant to the philosophical analysis of science. A consensus can be influenced or even forced by social, economic or political factors, e.g., the dominance of a particular individual or school, the availability of funds for a particular kind of research, or the prohibition of some kinds of research by a government or powerful religious organization. If scientific knowledge is determined by the existing consensus, rather than by an appeal to unambiguous, mechanically applicable standards, then it would seem impossible to distinguish a legitimately established consensus from an illegitimate one. Indeed, as long as we base our philosophy of science on the history of science, it would seem impossible to judge which methods of achieving consensus are acceptable and which unacceptable, for history can only tell us what has happened, not what ought to happen. Logical empiricists, on the other hand, took formal logic as the basis of their philosophy of science and logic is a normative discipline; it provides standards to which reasoning ought to conform irrespective of how reasoning may in fact have been carried out. Thus logic can provide the foundation for a normative philosophy of science while a historically based philosophy of science can only be descriptive.

Part of this argument is correct, although it is not accepted as a criticism: it does follow from our analysis that the traditional boundaries between what is and is not relevant to the philosophical analysis of science must be shifted. An example will help to make the point. Toulmin criticizes Kuhn for failing to distinguish between the influence of Newton's *Principia* and his *Optics*:

> Thus, while Newton's dynamical theories retained a legitimate intellectual authority of their own until the year 1880 or later, the influence of the *Optics* was already having a narrowing effect before the end of the eighteenth century. By 1800, in fact, the continued authority of the *Optics* represented little more than the magisterial dominance of a great mind over lesser ones, and the ways in which Newtonian scientists invoked this authority were beginning to lapse into dogmatism.[16]

Kuhn, according to Toulmin has failed to distinguish "a worthwhile philosophical point: namely that one function of an established conceptual scheme is to determine what patterns of theory are available, what questions are meaningful, and what interpretations are admissible," from "a different, sociological point: namely, that in historical fact secondary or derivative scientists, like the eighteenth century Newtonians, tend to see less of the whole picture than the primary, original workers who were their mentors, and who provided their inspiration."[17]

Now Toulmin's judgment of the relative merit of *Principia* and the *Optics* may be correct, but it is important to remember that this judgment is made in retrospect, not from the viewpoint of the scientists involved. From the latter viewpoint, the fact remains that research was done in the late eighteenth and early nineteenth centuries under the guidance of Newton's *Optics*, and this research constituted a part of the physics of that day. It might well be the case that important discoveries were postponed because of Newton's prestige, but these potential discoveries which had not yet been made were not a part of physics in 1800. Similarly, to take a contemporary example, experimental research in high energy physics is extremely expensive and can only be carried out in political and economic circumstances which allow for adequate funding.[18] Research that is not done does not provide information and thus does not contribute to scientific knowledge. Such facts are not relevant to the philosophical analysis of science for those who identify philosophical analysis with logical analysis. Our concern, however, is to understand how a consensus is established, and no factors which may enter into the establishment of this consensus can be rejected as irrelevant a priori.

At the same time, we do not accept the conclusion that a historically based philosophy of science cannot provide any norms which allow us to distinguish a legitimately established consensus from an illegitimate one. The main source of the criticism is the Humean doctrine[19] that we cannot derive "ought" from "is," that we cannot validly *deduce* normative statements about how science should function from a description of how it does function. But it is not necessary to provide a formal deduction in order to establish a norm. We have seen that fundamental scientific reasoning is dialectical, that new proposals grow out of existing theories without being formally entailed by these theories, and we have also seen that the decision to pursue a particular line of research is a matter of informed judgment on the part of scientists and not a matter of rote application of rules. A similar approach applies to philosophy of science. We can look back over the history of science and, judging from the present situation, attempt to discover what procedures have advanced the development of science and what have tended to retard it. On the basis of this information we can make recommendations about scientific procedure: not, to be sure, recommendations about what kinds of experiments scientists should carry out or what kinds of theories they should attempt to construct; these are judgments which must be left to the scientists themselves. But we can make recommendations of a more general sort which may well be, in the long run, more important than any decision to do one kind of experiment or computation rather than another.

As an initial example, consider the attempt by the Catholic Church during the seventeenth century to block the development and dissemination of Copernican astronomy. The most well-known episode in this attempt is the suppression of Galileo's *Dialogue Concerning the Two Chief World Systems* by the censors, and his subsequent arrest which resulted in a forced abjuration of Copernicanism and house arrest for the closing years of his life. Now we can recognize that there was a genuine scientific controversy raging in astronomy, but one in which theological issues were entangled, making inevitable the involvement of the Church. We can nevertheless condemn the Church's attempt to silence Galileo as the kind of approach that, were it successful, would tend to be destructive of science. As historians, we can study the political and intellectual structure of seventeenth century Italy and identify the forces that led to the attempt to stop Galileo's research. But although we can try to understand the dynamics of Galileo's trial from the standpoint of its actors, we need not evaluate it from that standpoint. Indeed, it would be absurd to do so, for we have a great deal of information about the trial, its aftermath, and the fruits of Galileo's refusal to cease advocating the Copernican system, and in order to evaluate the trial from a seventeenth century standpoint we would have to ignore all this information. In contrast to Hempel's argument that the paradoxes of confirmation are a psychological illusion resulting from the use of information not formally relevant to the problem at issue,[20] I have maintained that a rational judgment must be based on all available information even if the resulting judgment does not follow deductively from that information. It is on the basis of this information that we can condemn the attempt to silence Galileo as anti-science.

It might be objected that this argument begs the question since it takes the present scientific consensus as its starting point and rejects as unacceptable those means of achieving consensus which would have tended to prevent its development. But this criticism has force only on the assumption that it is possible to pass judgments from some presuppositionless vantage point; this is a philosophical presupposition which we have rejected. Evaluations must be made from some viewpoint and the only way in which we can evaluate the scientific significance of past disputes is from the perspective of present science. At the same time, of course, we judge present science in the light of the fate of past science and recognize that all of our judgments are fallible.

The inevitability of making some assumptions about what constitutes legitimate science in order to make philosophical judgments about science has been implicit in all work in the philosophy of science, whether overtly recognized or not. The strict verificationist theory of meaning, for example, which canonized as a philosophical principle what its advocates took to be a general feature of science, was abandoned by most positivists when it was realized that it would have eliminated just those segments of science which its proponents took to be paradigmatic of all knowledge. Even writers such as Schlick and Waismann who clung to strict verifiability did so by means of a new interpretation of the universal statements that the principle eliminated, one which dissolved the conflict between philosophical theory and the data to which it was addressed.[21] Does this too beg the question? After all, once a philosophical principle is accepted, must we not apply it uniformly in all relevant cases? But we must remember that it was reflection on science, not an insight into the nature of meaning, that gave rise to the principle of verifiability, and it was the continuing attempt to interpret science on this basis that led to the modification and eventual rejection of the principle. This is the kind of interplay that is characteristic of dialectic, and it is this kind of dialectical development over time and involving several participants that is characteristic of research.

Let us take another example. In discussing the genesis of his demarcation criterion Popper emphasizes that he sought a criterion which would eliminate certain theories he considered to be psuedo-sciences, astrology, for example, but more importantly, Marxism and Freudian and Adlerian psychology.[22] It is not surprising that Popper succeeded in finding a criterion which allowed him to arrive at these pre-conceived conclusions. Is he too arguing in a circle? If so, there does not seem to be any way in which we can avoid such circles. If we attempt to provide a definition of science by studying what the various sciences have in common, we must have some initial definition of "science" in order to know what disciplines to examine. And if we begin with an a priori definition, we may find that some of the most obvious examples of science have been eliminated, as strict falsificationism would eliminate Newtonian dynamics, since its proponents ignored clear counter-instances, and strict verificationism would eliminate all universal propositions. The way to avoid this dilemma is to reject the demand that we opt for one horn or the other. No one ever undertakes a philosophical analysis of science, or even a piece of scientific research, without a substantial body of prior commitments about how to proceed. If this generates a circle, it is not a vicious one since elements of the circle can be modified and the entire circle can be abandoned and replaced in the course of ongoing research.

With the above discussion as background we will examine two contemporary cases in which the problem of how a legitimate consensus is arrived at is particularly clear: the cases of Lysenko and Velikovsky. In both cases we illustrate how recommendations about what constitutes a legitimate approach to a scientific consensus can be made on the basis of case studies. We will begin with the Lysenko case in the Soviet Union.[23]

T. D. Lysenko was an agronomist who, beginning in the late 1920's and early '30's, developed a series of theoretical claims in biology and practical proposals for agriculture which entailed the rejection of modern genetics. In its place he substituted his "theory of the phasic development of plants" according to which it is possible, at certain phases in the life of a plant, to "shatter" its heredity and replace it with a new one, thus changing one species into another. With his collaborator, I. I. Prezent, Lysenko launched an attack on established biology in which his main weapon was political demogoguery, opponents being labeled advocates of "bourgeois science," "idealism," "trotskyism," and so forth. Through a variety of circumstances, including the structure of the Soviet political and industrial system at the time, success in gaining Stalin's favor,[24] and an agriculture crisis, Lysenko achieved complete control over Soviet biology.[25] Many opposing scientists were arrested, others lost their jobs, from 1948 to 1953 research in and the teaching of genetics were banned, and biology textbooks were rewritten in accordance with the "new biology." Lysenkoites continued to control the universities and scientific journals well into the '60's.

We would like to be able to say that the techniques used to silence the opposition and create a consensus in this case are not acceptable. Yet if scientific knowledge is the scientific consensus, then Lysenkoism might seem to be as good an example of science as any other theory that achieves a consensus, irrespective of the means used. But this conclusion is incorrect since there never was a Lysenkoite consensus. At most there was a consensus in one country, but the scientific community knows no national boundaries and the world biological community certainly never accepted Lysenko's theories. Indeed, there was never a Lysenkoite scientific consensus even in the Soviet Union. The opposition was, for a time, forced to go underground, but it remained strong with scientists in other disciplines helping keep genetics research alive under the guise of chemistry, physics and mathematics. Thus the interference of the Soviet government did not even succeed in eliminating the previously existing scientific consensus.

We have still, however, not faced the central issue posed by the Lysenko affair: the possibility of a forced consensus. Suppose this sort of repression were to occur on such a scale that a single orthodoxy is imposed on all scientists with all dissent and debate eliminated. Would we have a scientific consensus? To answer this question we must remember that the consensus view does not permit everyone to take part in developing a consensus. It is trained scientists who are the arbiters of scientific questions and a scientific consensus must be a consensus of the relevant scientific community. Not everyone has a right to an opinion on biology or physics or mathematics, any more than on brain surgery. Whatever semblance of a Lysenkoite consensus existed was brought about by the police powers of a totalitarian state, and thus was not a *scientific* consensus. If a sufficiently powerful police state did succeed in imposing a uniform view on all, it would have eliminated a scientific discipline or perhaps all science, but it would not have brought about a scientific consensus.

There is another direction from which we may be charged with evading the problem rather than solving it. Having rejected the view that fundamental scientific decisions are made by algorithms and shifted the burden onto the scientific community, we have not, it may be argued, provided any clear criteria for who is to count as a member of this community. But again we must remember what epistemology we are working with. The objection, in effect, asks for an algorithm which will allow us to determine the composition of the relevant scientific community, and there does not seem to be any more prospect of constructing one than there is of constructing an algorithm for deciding between fundamental theories. But, once again, this does not imply that these decisions are irrational. Scientific communities developed because numerous researchers worked on common problems, and they are perpetuated by the process of educating new researchers. This process is fraught with the risks of social life: of discipleship, of ignoring the "oddball" innovator, of placing dull conforming scientists in authority. But these risks are to be expected since no community is infallible in any of its decisions. Yet one of the most intriguing aspects of science is its self-correcting ability. The intellectual damage done by human fallibility and perversity has often been relatively short-lived with many proposals that were once rejected getting a new hearing later, and some of these eventually being accepted.[26]

A dramatic illustration of the dangers inherent in the consensus process is provided by the Velikovsky affair. In 1950 Immanuel Velikovsky published *Worlds in Collision* in which he challenged a number of widely accepted theses in astronomy and cosmology.

He tried to show that the earth has undergone cataclysmic changes in historical times as a result of collisions or near collisions with extra-terrestrial objects, especially a large comet which subsequently became the planet Venus, and later with Mars. Velikovsky's evidence was derived from the study of myths which, he maintained, are based on actual events. He compared myths of many cultures from all over the world and claimed to have found synchronous accounts of such catastrophies. As a result, he proposed a view of the nature of the solar system radically different from that of accepted science, and some of the conclusions he drew have since been verified, e.g., that there are strong electromagnetic fields in interplanetary space, that the sun carries a large electric charge, that the surface temperature of Venus is higher than was then believed, and that Jupiter emits radio noise. Here we will not attempt to assess Velikovsky's methodology—the way he has interpreted his sources, whether his conclusions follow from his evidence, etc.—nor will we compare his explanation of the new discoveries with others that have been proposed. What does concern us is the initial response of the scientific community to his book.

We have seen that it is common for scientists to ignore work that is inconsistent with currently accepted presuppositions, and that to a large degree this is legitimate and necessary. As Polanyi points out, "to drop one's work in order to test Velikovsky's claims, as requested by him, would appear a culpable waste of time, expense, and effort."[27] But Velikovsky was not ignored. Many scientists, including some who had refused to make observations he had requested, or even to read his manuscript, did drop their work in order to attack him. This raises serious questions about the way in which a scientific consensus is achieved and maintained; it will be useful to examine some of the events involved in this case.

There must be some way of filtering out genuine science from the work of incompetents and cranks, and the referee system of journals and publishing houses is responsible for this. Before publication Velikovsky's book was subjected to an unusually stringent referee process. It was originally recommended for publication by a Macmillan editor "an optional contract was signed, and then, after another year in which various outside readers—among them [John] O'Neill [science editor of the New York Herald Tribune] and Gordon Atwater, then Curator of the Hayden Planetarium and Chairman of the Department of Astronomy of the American Museum of Natural History—examined the manuscript and recommended publication, a final contract was drawn and signed."[28] But in 1950, in response to two letters from Harvard astronomer Harlow Shapley, who had previously refused to read the manuscript,[29] Macmillan decided to submit the book to three more readers and accept the majority decision. "Apparently the majority again voted thumbs up; the book was published on schedule."[30]

Unfortunately, Velikovsky and Macmillan permitted a number of rather sensational pre-publication articles on the book in popular magazines: one in *Harper's,* two in *Collier's,* and one in *Reader's Digest.*[31] It is the magazine articles, not the book, that became the focal point for the initial attacks by scientists on Velikovsky. The first, and most important of these attacks, since it was frequently referred to by later writers, was by astronomer Cecilia Payne-Gaposchkin, a colleague of Shapley's.[32] Dr. Gaposchkin's article appeared before the publication of *Worlds in Collision* and was based entirely on Larabee's article in *Harper's.* Indeed, as Gaposchkin acknowledged in response to a letter by Larabee,[33] she had not read the book when she wrote the article. Her attack was based on "an eight page summary obviously written by a non-scientist. . . ."[34] But this did not stop her from writing as if she were quoting directly from Velikovsky. Consider, for example, the following:

> Let us examine some of Dr. Velikovsky's astronomical assertions in detail. 'A comet . . . *did* pass close to the earth. . . . The comet . . . touched the earth with its gaseous tail . . . and with the shower of meteorites the earth stopped turning.'[35]

The quotation is not from Velikovsky, but from Larabee, and Gaposchkin's use of ellipses is, to be generous, misleading. For example, the sentence ending before the second ellipses occurs at the top of page 20 in Larabee's article, the passage beginning after the ellipses occurs at the top of page 21, a full page later, and in a different numbered section of the article.

The attacks, many of them quite intemperate, continued after the publication of the book, and a number of professional journals which printed these attacks refused to allow Velikovsky space for rebuttal. To take one example, in 1952 the American Philosophical Society held a symposium on "Some Unorthodoxies in Modern Science" which included attacks on Velikovsky. He was present at the meeting and the chairman permitted him to respond, but the Society refused to print his remarks in the *Proceedings* although the remarks of his opponents were printed.[36]

Perhaps the most serious aspect of the entire affair was a concerted, possibly an organized, effort to force the publisher to take the book out of print. Scientists wrote numerous angry letters to Macmillan, and refused to see their textbook salesmen or to write textbooks for them. The attack was so strong that they were forced to give up the book, even though it was their most profitable, assigning the rights to Doubleday, which had no textbook division.[37]

Finally, James Putnam, the editor who had recommended the book for publication, and who had been with Macmillan for 25 years, was fired[38] and Gordon Atwater was dismissed from his positions at the Hayden Planetarium and American Museum of Natural History after scheduling a planetarium program on Velikovsky's theories, the program being cancelled.[39]

We do not need a formal deduction, the probability calculus, or a solution to the is-ought problem to recognize that attempts to suppress a book and attacks on an author one has not read are not legitimate means of maintaining a scientific consensus. The main argument scientists used to justify the attempted suppression of the book is well stated by astrophysicist John Q. Stewart:

> When the Macmillan Company accepted this book for publication that house could not have been mindful of their own long list of works by leading investigators in mathematical physics and celestial mechanics. Readers unfamiliar with an author in physical science frequently take a respected publisher's imprint as something of a guarantee. This is particularly important when an expert in search of reliable information steps out of his own professional field into a neighboring one. . . . Free speech would not have been involved; publishers reject books every day. . . .[40]

Stewart's point that we often depend on the reliability of a publisher when we make use of a technical book is an important one.[41] But it is difficult to understand how experts from a neighboring scientific discipline could possibly mistake *Worlds in Collision* for a definitive work in astrophysics, and if they were misled, Macmillan's reputation should indeed have suffered.

More importantly, the book was accepted for publication only after it had been approved by a number of scientifically competent readers; there would have been no violation of free speech if these readers had rejected the book. But Shapley's attempt to interfere with the normal referee process and the attempt to suppress a book which has already been published are examples of a kind of activity which can only hurt science. The well-being of science depends on its openness for new ideas and on the maintenance of free debate. Once again, this does not require that scientists spend their professional lives examining every implausible conjecture that gets published just because some rejected theses later turn out to be important innovations. But it does mean that scientists who choose to enter a debate should behave in accordance with the normal standards of reasoned discussion. The fact that a writer proposes a theory in a field in which he is not a trained professional may provide an adequate reason for ignoring him, it does not provide a license for suspending the standards of scientific debate. It is useful to recall that throughout history science has been one of the major victims of censorship and there is no reason to believe that censorship becomes desirable when scientists are the censors. Rather, what the consensus principle teaches is that it is the process of continuing research and discussion that leads to the establishment of scientific knowledge. As Moses Hadas argued with respect to Velikovsky, "What bothered me was the violence of the attack on him: if his theories were absurd, would they not have been exposed as such in time without a campaign of vilification?"[42]

Here, as in so many other cases, a model of appropriate behavior is provided by Einstein. During the last two years of his life Einstein debated Velikovsky's theories with him, rejecting particularly his assertions that electromagnetic fields play a role in planetary motion and that the sun and planets carry an electric charge. "Yet when he learned, only days before his death, that Jupiter emits radio noise, as Velikovsky had long insisted, he offered to use his influence in arranging for certain other experiments Velikovsky had suggested."[43]

The violence of the reaction to *Worlds in Collision* is partly a result of the lingering belief that science produces final truths. Consider, for example, Shapley's remark, "If Dr. Velikovsky is right, the rest of us are crazy."[44] Clearly, if Velikovsky is right, a lot of astronomers are mistaken, but to equate holding an incorrect scientific theory with being crazy is to adhere to a naively optimistic view of the permanence of scientific results, especially in a field as speculative as cosmology. Yet even the most speculative cosmologists seem to have no hesitation in declaring the finality of their pronouncements. Fred Hoyle, for example, an advocate of steady-state cosmology, which rejects the principle of conservation of matter and postulates the continuous creation of hydrogen atoms from nothing in interstellar space, wrote in 1960:

> Is it likely that any astonishing developments are lying in wait for us? Is it possible that the cosmology of 500 years hence will extend as far beyond our present beliefs as our cosmology goes beyond that of Newton? It may surprise you to hear that I doubt whether this will be so. If this should appear presumptuous to you, I think you should consider what I said earlier about the observable region of the Universe. As you will remember, even with a perfect telescope we could penetrate only about twice as far into space as the new telescope at Palomar. This means that there are no new fields to be opened up by the telescopes of the future, and this is a point of no small importance in our cosmology. There will be many advances in the detailed understanding of matters that still baffle us. Of the larger issues I expect a considerable improvement in the theory of the expanding Universe. Continuous creation I expect to play an important role in the theories of the future. Indeed, I expect that much will be learned about continuous creation, especially in its connection with atomic physics. But by and large I think that our present picture will turn out to bear an appreciable resemblance to the cosomologies of the future.[45]

While Hoyle was writing these words, the recently developed field of radio astronomy was providing new kinds of data and forcing many cosmologists to rethink their theories. Thus in a 1965 book Hoyle found it necessary to include a chapter entitled "A Radical Departure from the Steady-State Concept" in which he attempted to reconcile steady-state theory with new evidence that the observable portion of the universe lacks the homogeneity required by that cosmology. The part of the universe we live in is, he argued, a local inhomogeneity "some 10^{13} light years in diameter, about a *thousand times the portion of the universe visible in our telescopes.*"[46] Yet, having been shown, within five years, to have been wrong on a number of fundamental questions. Hoyle still held out the hope that his latest theory was the final word: "My impression is that the picture . . . of the formation of the galaxies given by this theory may be decisive."[47]

Even a cursory study of twentieth century science makes it extremely difficult to understand such serene confidence, and it is equally difficult to see how science can benefit from perpetuating the myth that it is the "established results," rather than the ongoing research, that is the life-blood of science. The scientific community as a whole must maintain a delicate balance between accepted principles and new ideas,[48] and this balance can best be maintained if scientists have a clear understanding of how science has developed.[49]

It is not my purpose here to present a catalog of norms for scientists, but only to illustrate the way in which such norms can be proposed on the basis of analysis of cases from the history of science. These norms cannot be formally deduced from descriptions of historical events, nor is their any a priori proof that they ought to be followed, but there are no good reasons for making either of these demands necessary conditions for the acceptability of proposed norms. It is sufficient that we be able to recognize the kinds of behavior that have tended to help or hinder the development of science in order to make proposals as to how scientists ought to behave. Thus the question "Is the philosophy of science normative or descriptive?" is not particularly illuminating if it presupposes that these are mutually exclusive alternatives.

Presuppositions
and Problems

I have argued that a theory is a system of presuppositions which guides research and that the epistemology developed here is an example of such a theory. What, then, are the presuppositions of this theory, and what research problems does it provide for philosophers? These are not two distinct questions in our case, for the clarification of presuppositions is a philosophical problem whether or not the theory in question is itself a philosophical theory. In general it is difficult for those working within a theory to be fully clear about its presuppositions, but we can attempt to formulate the more obvious ones. In doing so we will find that each presupposition we isolate will generate further philosophical problems.

Our analysis of science has made extensive use of the history of science and we have presupposed throughout that we have adequate historical information with which to work; but we must not assume (on pain of self-contradiction) that historical research provides theory-free data. Thus our reliance upon the history of science requires a theory of historical knowledge within the context of our epistemology and the reassessment of our analysis of science in the light of this theory of history.

Second, we have presupposed that there exists a mind-independent material world which is the object of the scientist's theories and which plays a crucial role in determining what is observed. This requires a new theory of perception which can clarify the roles that theory and physical reality play in determining what we observe, and how such theory-laden observation can serve to control the acceptance and rejection of scientific theories.

Third, we have presupposed that the human mind is capable of rational judgment on the basis of limited data and without the guidance of effective procedures. This presupposition (which is rejected by those who equate rationality with algorithmic computability) necessitates a new epistemological theory of mind to account for scientific rationality and creativity in forming and assessing hypotheses.

Fourth, a new field for philosophical analysis of the nature of knowledge is opened up. Just as our increasing understanding of the mechanisms of evolution and genetics have raised the possibility of altering these, an understanding of the role of presuppositions in human knowledge may also lead to the possibility of altering their role, and this, in turn, would generate the need for another new epistemology.

Conclusion

Our central theme has been that it is ongoing research, rather than established results, that constitutes the life-blood of science. Science consists of a sequence of research projects structured by accepted presuppositions which determine what observations are to be made, how they are to be interpreted, what phenomena are problematic, and how these problems are to be dealt with. When the presuppositions of a scientific discipline change, both the structure of that discipline and the scientist's picture of reality are changed. The only permanent aspect of science is research.

This approach has been applied to philosophy of science as well as to science. We have seen that logical empiricism is an attempt to interpret science in terms of an accepted philosophy, using a definite body of intellectual tools, and that it generates a characteristic set of problems which philosophers working in this tradition attempt to solve. Failures to solve these problems led to modifications of the original research program, and eventually to the proposal of a new philosophy of science built on different presuppositions, using different intellectual tools, and generating an alternative set of philosophical problems. At the same time, the boundary between what is and is not relevant to the philosophical analysis of science is shifted, with many aspects of the history, sociology, psychology, and even the economics and politics of science, which are considered irrelevant by those who identify philosophy of science with formal analysis, becoming very relevant from the new standpoint.

The attempt to develop a philosophical theory of science requires that we change our interpretation of science when we alter our epistemological presuppositions, and that we change our presuppositions when the problems about science that they generate are judged intractable. This continuing interplay between a theory and its object has been characterized as dialectical, and it has been argued that the structure of scientific research and development is also dialectical. It consists of a transaction between theory and observation, where theory determines what observations are worth making and how they are to be understood, and observation provides challenges to accepted theoretical structures. The continuing attempt to produce a coherently organized body of theory and observation is the driving force of research, and the prolonged failure of specific research projects leads to scientific revolutions.

Scientific revolutions are not, however, sharp breaks with tradition which institute a new approach having nothing in common with earlier science. We have seen that in introducing new presuppositions, a revolution transforms the conceptual structure of a theory. This may involve the elimination of some concepts and the rejection of some forms of observation as irrelevant, along with the introduction of some new concepts and new kinds of observations. But, for the most part, old concepts are retained in altered form and old observations are retained with new meanings. This continuity provides the basis for rational debate between alternative fundamental theories, even though these theories may present radically different pictures of nature and of the discipline in question. Thus the thesis that a scientific revolution requires a restructuring of experience akin to a gestalt shift is compatible with the continuity of science and the rationality of scientific debate.

Finally, it has been argued that such crucial decisions as how a conflict between theory and observation is to be resolved, or how a proposed new theory is to be evaluated, are not made by the application of mechanical rules, but by reasoned judgments on the part of scientists and through debate within the scientific community. This, admittedly fallible, process is offered as a paradigm of a rational decision procedure.

Notes

Chapter One

1 David Hume, *A Treatise of Human Nature,* ed. L. A. Selby-Bigge (Oxford U. Press, 1967), p. 1.

2 *Ibid.,* pp. 70-71

3 David Hume, *An Enquiry Concerning Human Understanding,* ed. L. A. Selby-Bigge (Oxford U. Press, 1966), p. 25.

4 Bertrand Russell, *Principles of Mathematics,* second edition (W. W. Norton, 1937), p. xv.

5 Alfred North Whitehead and Bertrand Russell, *Principia Mathematica* (Cambridge U. Press, 1962), p. 94.

6 Bertrand Russell, *Introduction to Mathematical Philosophy* (George Allen and Unwin, 1919), p. 153.

7 *Ibid.,* pp. 204-5.

8 *Ibid.,* p. 205.

9 Ludwig Wittgenstein, *Tractatus Logico-Philosophicus,* trans. D. F. Pears and B. F. McGuinness (Routledge and Kegan Paul, 1961), I. I. (In all references to the *Tractatus* proposition numbers, rather than page numbers, will be given.)

10 *Ibid.,* 1.2.

11 *Ibid.,* 1.21.

12 *Ibid.,* 6.37.

13 *Ibid.,* 4.21.

14 *Ibid.,* 4.211.

15 *Ibid.,* 1.2.

16 *Ibid.,* 2.01.

17 Rudolph Carnap, "Testability and Meaning," *Philosophy of Science,* 3, 1936, pp. 419-471 and 4, 1937, pp. 1-40. Reprinted

with omissions in *Readings in the Philosophy of Science,* eds. Feigl and Brodbeck (Appleton, Century, Crofts, 1953), pp. 47-92. All further references to this essay are to the reprinted edition.

18 *Ibid.,* p. 48.

19 *Ibid.,* pp. 63-65.

Chapter
Two

1 Alternatively, we can construct an abstract calculus which we intend to interpret as a calculus of confirmation, but in order to carry out this interpretation with respect to some set of statements we must have already distinguished those statements into hypotheses and statements of confirming or disconfirming instances.

2 Carl G. Hempel, "Studies in the Logic of Confirmation," *Mind,* 54, 1945, pp. 1-26 and 97-121. Reprinted with a postscript in Carl G. Hempel, *Aspects of Scientific Explanation* (Free Press, 1965), pp. 3-51 (hereafter referred to as *Aspects*); further references will be to the *Aspects* version. Cf. Carl Hempel, "A Purely Syntactical Definition of Confirmation," *Journal of Symbolic Logic,* 8, 1943, pp. 122-143.

3 *Aspects,* p. 10.

4 The term "research program" is taken over from the work of Imre Lakatos, although I do not make use of his analysis of the structure of research programs. Cf. "Falsification and the Methodology of Research Programmes," *Criticism and the Growth of Knowledge,* ed. I. Lakatos and A. Musgrave (Cambridge U. Press, 1970), pp. 91-195.

5 *Aspects,* pp. 10-11. Among the difficulties Hempel mentions in his discussion of Nicod's criterion is that this cannot be a completely general characterization of the confirmation relation since it holds only for universally general propositions (*Ibid.,* pp. 11-12). We will ignore this point here (as does Hempel throughout most of his discussion) and limit ourselves to the confirmation of universal propositions.

6 *Ibid.,* p. 12.

7 *Ibid.*

8 *Ibid.,* p. 13.

9 *Ibid.,* p. 15.

10 Nelson Goodman, *Fact, Fiction and Forecast* (Bobbs-Merrill, 1965), pp. 70-71.

11 R. G. Swinburne, "The Paradoxes of Confirmation—A Survey," *American Philosophical Quarterly,* 8, 1971, pp. 318-330.

12 *Aspects,* p. 18.

13 *Ibid.,* pp. 18-19.

14 *Ibid.,* pp. 19-20.

15 *Ibid.,* p. 20.

16 *Ibid.*

17 This claim will be defended at length in Part II.

18 Richard B. Angell, "The Boolean Interpretation is Wrong,"

171   *Readings on Logic,* Irving M. Copi and James A. Gould, eds. (Macmillan, 1972), p. 182.

19 *Aspects,* p. 31. This is a "special consequence condition" in comparison to what Hempel calls the *consequence condition*: "If an observation report confirms every one of a class *K* of sentences, then it also confirms any sentence which is a logical consequence of *K*." *Ibid.*

20 *Ibid.,* p. 32.

21 William Barrett, "On Dewey's Logic," *Philosophical Review,* 50, 1941, p. 312.

22 *Aspects,* p. 32.

23 Rudolph Carnap, *The Logical Foundations of Probability,* 2d ed. (U. of Chicago Press, 1962), pp. 474-5.

24 Goodman, *Fact, Fiction and Forecast,* p. 74.

25 *Ibid.,* p. 81.

26 *Ibid.,* pp. 57-8.

27 *Ibid.,* pp. 84-5.

28 Cf. also *Ibid.,* pp. 70 and 75.

29 *Ibid.,* p. 94.

30 *Ibid.*

31 Goodman foresaw this line of criticism and included a response in his original argument: *Ibid.,* pp. 79-80.

Chapter Three

1 In introducing the term "sense-data" as a substitute for Hume's "impressions" I am, in effect, eliminating from consideration the meaning of terms of introspective psychology which were included in Hume's analysis. This is wholly consistent with contemporary empiricism.

2 Bertrand Russell, "The Relation of Sense-Data to Physics," *Mysticism and Logic* (Doubleday Anchor Books, 1957), p. 150. It is worth noting that Russell finds it necessary to allow himself two inferred entities, the sense-data of other people and what he calls "sensibilia," i.e., the sense-data that would appear if there were an observer at a position at which there does not happen to be one at present. *Ibid.,* p. 152.

3 *Ibid.,* p. 151.

4 F. P. Ramsey, "Theories," *The Foundations of Mathematics* (Littlefield, Adams, 1960), pp. 212-236.

5 R. B. Braithwaite, *Scientific Explanation* (Harper Torchbooks, 1960), Chap. III.

6 Cf., for example, Israel Scheffler, *Science and Subjectivity* (Bobbs-Merrill, 1967), pp. 39-41 and Ch. 3, and Carl R. Kordig, *The Justification of Scientific Change* (Humanities Press, 1971), Ch. 2 and 3.

7 P. W. Bridgman, *The Logic of Modern Physics* (Macmillan, 1927), p. 5.

8 *Ibid.*

9 *Ibid.,* p. 10.

10 *Ibid.,* pp. 14-23.

11 "Testability and Meaning," pp. 52-3.

12 Some twenty years later Carnap wrote of "Testability and Meaning," "At the time of that paper, I still believed that all scientific terms could be introduced as disposition terms on the basis of observation terms either by explicit definitions or by so-called reduction sentences, which constitute a kind of conditional definition." "The Methodological Character of Theoretical Concepts," *Minnesota Studies in the Philosophy of Science,* I, ed. Herbert Feigl and Michael Scriven (U. of Minnesota Press, 1956), p. 53. As the quoted passage suggests, by 1956 Carnap no longer held that theoretical terms could be treated as disposition terms. *Ibid.,* pp. 66-69.

13 It must be kept in mind that $P, Q, R, S,$ and T in the above sentences are all predicates, so that strictly speaking the reduction sentences should be quantified expressions. (1), for example, should read: (x)(t)[Pxt (Qxt Rxt)] but I will follow Carnap's practice of suppressing the quantifiers since it should cause no confusion in the present context.

14 "Testability and Meaning," p. 63.

15 *Ibid.,* p. 53.

16 Henry S. Leonard, "Review of Rudolph Carnap, 'Testability and Meaning'," *Journal of Symbolic Logic,* 2, 1937, p. 50.

17 *Fact, Fiction and Forecast,* p. 47.

18 Carl G. Hempel, *Fundamentals of Concept Formation in Empirical Science,* (U. of Chicago Press, 1952), p. 24.

19 *Ibid.,* p. 23.

20 *Ibid.,* pp. 28-29.

21 "Testability and Meaning," p. 60.

22 *Ibid.,* p. 61.

23 Carl G. Hempel, "Empiricist Criteria of Cognitive Significance: Problems and Changes," *Aspects,* pp. 114-115.

24 The deduction is as follows:

$$P \supset (Q \equiv R) \qquad\qquad A \supset (B \equiv R)$$
$$P \supset (Q \supset R) \qquad\qquad A \supset (R \supset B)$$
$$(P \cdot Q) \supset R \quad (1) \qquad\qquad R \supset (A \supset B) \quad (2)$$

from (1) and (2)
$$(P \cdot Q) \supset (A \supset B) \text{ or}$$
$$(P \cdot Q \cdot A) \supset B$$

25 Cf. *Fundamentals of Concept Formation,* p. 80, n. 21.

26 William Craig, "On Axiomatizability Within a System," *Journal of Symbolic Logic,* 18, 1953, pp. 30-32. See also William Craig, "Replacement of Auxiliary Expressions," *Philosophical Review,* 65, 1956, pp. 38-55. For discussions of the relevance of this theorem to the problem of theoretical concepts in science, see Hempel, *Aspects,* 210-215; Ernest Nagel, *The Structure of Science* (Harcourt, Brace, World, 1961), pp. 134-137; Israel Scheffler, *The Anatomy of Inquiry* (Bobbs-Merrill, 1963), pp. 193-203.

27 "Replacement of Auxiliary Expressions," p. 39.

28 *Ibid.,* p. 49.

29 *Ibid.,* p. 41.

30 Craig mentions that it was Hempel who encouraged and advised him to write a nontechnical exposition of the theorem. *Ibid.*, p. 39.

31 Another attempt to eliminate theoretical terms has been based on a formal device introduced by Ramsey ("Theories"), but a detailed discussion will not add to the points already made. Suffice it to say that empiricists who have discussed it have not found it any more satisfactory than Craig's method. See Hempel, *Aspects,* pp. 215-217 and Scheffler, *Anatomy of Inquiry,* pp. 203-222.

32 Norman Robert Campbell, *Physics: The Elements,* 1920, reprinted as *Foundations of Science* (Dover, 1957), pp. 122-129. References are to the reprinted edition.

33 *Ibid.*, pp. 123-4. Campbell called the two parts of a theory the "hypothesis" and the "dictionary" (*Ibid.*, p. 122), but I will continue to use the current term "correspondence rules" for the latter.

34 Herbert Feigl, "The 'Orthodox' View of Theories," *Minnesota Studies in the Philosophy of Science,* IV, ed. Michael Radner and Stephen Winokur (U. of Minnesota Press, 1970), p. 6.

35 *Ibid.*, p. 5.

36 *Ibid.*, p. 6.

37 *Ibid.*, p. 7. Similarly, Braithwaite describes the relation between observation terms and theoretical terms as "like a zip-fastener . . ." *Scientific Explanation,* p. 51. It is well worth noting just how far we have moved from the original project, which rejected as meaningless any term which could not be given a precise, explicit definition. In its latest versions the empiricist theory of meaning has itself been reduced to a metaphor.

38 Carl G. Hempel, "On the 'Standard Conception' of Scientific Theories," *Minnesota Studies,* IV, p. 159.

39 *Ibid.*, pp. 159-160.

40 *Ibid.*, pp. 143-144.

41 Cf. Dudley Shapere, "Notes Toward a Post-Positivistic Interpretation of Science," *The Legacy of Logical Positivism,* ed. Peter Achinstein and Stephen F. Barker (Johns Hopkins U. Press, 1969), p. 126: "It seems reasonable to ask whether some problems that have arisen in the context of discussions relying on the theoretical-observational distinction are *created, at least in part, by the limitations of that technical distinction and the roles for which it was introduced. If so, then the problematic character of those created 'problems' must be reconsidered in the light of the failures of the background against which they arose.*" However, I must disagree with Shapere's suggestion that such problems are "artificial" (*Ibid.*) because they were generated by the acceptance of a technical distinction. This distinction is an expression of a central thesis of empiricist epistemology and the problems that it generates are genuine problems which arise from the decision to analyze science from the point of view of that epistemology.

42 Hempel, "On the 'Standard Conception' of Scientific Theories," p. 162. Cf. Hilary Putnam, "What Theories Are Not," *Logic, Methodology and Philosophy of Science,* ed. Ernest Nagel, Patrick Suppes, Alfred Tarski (Stanford U. Press, 1962), p. 241.

43 "On the 'Standard Conception' of Scientific Theories," p. 163.

Chapter
Four

1 For the moment I am neglecting statistical explanation, although when we discuss this topic we will see that the statement that the deductive model is taken as basic still holds in an important sense.

2 Carl G. Hempel and Paul Oppenheim, "Studies in the Logic of Explanation," *Philosophy of Science,* 15, 1948, pp. 135-175. Reprinted with some changes and a postscript by Hempel in *Aspects,* pp. 245-295. References are to the reprinted edition.

3 *Ibid.,* pp. 248-249.

4 Hempel later modified his position on this point by distinguishing between true explanations and more or less well confirmed explanations. Cf. "Deductive-Nomological vs. Statistical Explanation," *Minnesota Studies in the Philosophy of Science,* III, ed. Herbert Feigl and Grover Maxwell (U. of Minnesota Press, 1962), p. 102 and "Aspects of Scientific Explanation," *Aspects,* p. 336.

6 *Aspects,* pp. 276-278.

7 *Ibid.,* p. 277.

8 Subsequent discussion has shown that this will not completely solve the problem since it remains possible, by the use of somewhat more complex symbolic maneuvers, to construct other anomalous results of the same sort, and, of course, to propose other, somewhat more complex means of removing these further anomalies. See *Aspects,* pp. 293-295. It will not advance our discussion here to consider the details of this further debate, but it is important to understand why the carrying on of a discussion of this sort is taken to be a contribution to the philosophy of science, i.e., to understand the discussion in terms of the logical empiricist presuppositional framework.

9 *Aspects,* p. 249.

10 *Ibid.,* p. 277.

11 *Ibid.,* p. 249.

12 The following argument holds a *fortiori* if only well confirmed premises are required.

13 Cf., for example, Campbell, *Foundations of Science,* pp. 129-132; Mary B. Hesse, *Models and Analogies in Science* (U. of Notre Dame Press, 1970).

14 *Aspects,* p. 445.

15 *Ibid.,* p. 439.

16 *Ibid.,* pp. 440-441.

17 Rom Harré, *The Principles of Scientific Thought* (U. of Chicago Press, 1970), p. 15. Harré proposes to describe a complete revolution in the philosophy of science, according to which models are essential to theories and the creation of deductive systems has only heuristic value.

18 I am not suggesting that this is an adequate solution to the problem, but only that it is the solution which is implicitly used. It should be noted that Hempel, in his discussion of Campbell, does propose a solution: "a worthwhile scientific theory explains an empirical law by exhibiting it as one aspect of more comprehensive underlying regularities, which have a variety of other testable aspects as well, i.e., which also imply various other empirical laws." *Aspects*, p. 444. But this proposal is completely consistent with the prediction criterion and still must be supplemented by the prediction criterion in order to avoid the very problem we have been discussing.

19 *Aspects*, p. 367.

20 Michael Scriven, "Explanation and Prediction in Evolutionary Theory," *Science,* 130, 1959, p. 480.

21 *Aspects*, pp. 369-370.

22 Michael Scriven, "Explanations, Predictions, and Laws," *Minnesota Studies,* III, pp. 192-193.

23 *Ibid.*, p. 205.

24 May Brodbeck, "Explanation, Prediction and 'Imperfect' Knowledge," *Minnesota Studies,* III, pp. 231-272.

25 *Ibid.*, p. 250.

26 *Anatomy of Inquiry*, p. 35.

27 *Ibid.*

28 *Ibid.*, pp. 40-42.

29 The question of whether explanation must be deductive is important for the practice of science as well as for the philosophy of science. A scientist who holds that scientific explanations must be deductive will take a case of non-deductive explanation such as Jones' heart attack as an occasion for further research, while one who holds that statistical explanations are sufficient may well conclude that there is no further work to be done here. This is, of course, essentially the kind of situation that is involved in the refusal of some contemporary physicists to accept the statistical physics provided by quantum theory as complete.

30 *Aspects*, p. 383.

31 Cf. "It is disquieting that we should be able to say: No matter whether we are informed that the event in question . . . did occur or that it did not occur, we can produce an explanation of the reported outcome in either case; and an explanation, moreover, whose premises are scientifically established statements that confer a high logical probability on the reported outcome." Hempel, *Aspects*, p. 396. Hempel has discussed this problem at length in a number of publications: for example, "Inductive Inconsistencies," *Aspects*, pp. 53-67; "Deductive-Nomological vs. Statistical Explanation," pp. 128-149; "Aspects of Scientific Explanation," *Aspects*, pp. 394-403. Cf. S. F. Barker, *Induction and Hypothesis* (Cornell U. Press, 1957), pp. 75-78.

32 *Aspects*, p. 393.

33 *Ibid.*, p. 58.

34 *Ibid.*, p. 395. There is at least one other reason why the ambiguity should be taken as problematic: the result of the two explanations is that both E and $\sim$E have a high probability while the sum of these probabilities must be one. Cf. Hempel, "Deductive-Nomological

vs. Statistical Explanation," p. 140. But it should also be noted that Hempel only mentions this aspect of the problem in passing in his 1962 essay and it plays no role in his 1965 discussion (*Aspects*, pp. 380-412). It seems quite clear that it is the comparison with deduction which is Hempel's major concern.

35 Cf. note 1 of this chapter.

36 *Logical Foundations of Probability,* p. 211. Cf. also Rudolph Carnap, "On the Application of Inductive Logic," *Philosophy and Phenomenological Research,* 8, 1947, pp. 138-139.

37 "Deductive-Nomological vs. Statistical Explanation," p. 124.

38 *Ibid.*

39 Ernest Nagel, "The Meaning of Reduction in the Natural Sciences," *Philosophy of Science,* ed. Arthur Danto and Sidney Morgenbesser (Meridian Books, 1960), pp. 290-291. Cf. also Hans Reichenbach, *The Rise of Scientific Philosophy* (U. of California Press, 1966), p. 101. Reichenbach does not even mention the need for boundary conditions.

40 The relation between Newtonian mechanics and the theory of relativity will be examined at length in chapters VIII and IX.

41 Cf. Karl Popper, "The Aim of Science," *Objective Knowledge* (Oxford U. Press, 1972), pp. 200-201.

42 *Aspects,* p. 344. See also "Deductive-Nomological vs. Statistical Explanation," pp. 100-101, for an earlier approach to this position. There Hempel still asserts in the text that Galileo's law can be deduced from Newton's but mentions that this is not correct in a footnote. In the later text that I have quoted, he gives the correct relation between the two laws but introduces the notion of an "approximative D-N explanation" in attempting to handle the anomaly while making the minimal change in the presuppositional base.

43 "Deductive-Nomological vs. Statistical Explanation," p. 108.

44 The attempts of Hempel and Nagel to save the laws of Kepler and Galileo from refutation will, of course, serve as the first illustration of this point; the extended discussion of the relation between Newtonian and relativistic dynamics in chapter VIII will serve as a second.

45 Hans Reichenbach, *Experience and Prediction* (U. of Chicago Press, 1938), p. vi.

46 *Ibid.,* p. 16. Italics mine.

47 *Rise of Scientific Philosophy,* p. 170.

48 *Ibid.,* p. 173.

49 *Ibid.*

50 *Aspects,* p. 248.

51 *Ibid.,* pp. 248-249.

52 "Deductive-Nomological vs. Statistical Explanation," p. 120; *Aspects,* p. 338.

53 *The Structure of Science,* p. 43.

54 *Ibid.*

55 Herbert Feigl, "Beyond Peaceful Coexistence," *Minnesota Studies,* V, 1970, ed. Robert H. Stuewer, p. 9.

56 *Ibid.,* p. 3.

1 Karl R. Popper, *The Logic of Scientific Discovery* (Harper Torchbooks, 1968), originally published in 1935 as *Logik der Forschung*. References are to the English translation which will be referred to as *LSD*.

2 See, for example, Alfred Jules Ayer, *Language, Truth and Logic* (Dover, 1946), p. 38; Reichenbach, *Experience and Prediction*, p. 88.

3 See below, p. 87.

4 *LSD*, p. 34.

5 *Ibid.*, pp. 35-37. Cf. also p. 40 n. *3. The asterisk indicates that the note was added on the English edition of 1954.

6 *Ibid.* pp. 28-30.

7 *Ibid.*, p. 33.

8 *Ibid.*, pp. 40-41.

9 *Ibid.*, p. 268.

10 Karl R. Popper, *Conjectures and Refutations* (Harper Torchbooks, 1968), p. 36.

11 *Ibid.*, p. 38.

12 *LSD*, p. 69.

13 *Ibid.*, pp. 112-113.

14 *Ibid.*, pp. 270-272.

15 *Ibid.*, p. 279.

16 *Ibid.*, pp. 265-269.

17 *Ibid.*, p. 268. More recently Popper has changed his position on this question and attempted to construct a formula for computing numerical degrees of corroboration, *Ibid.*, Appendix ix. According to his formula the degree of corroboration is a function of probabilities but it is not itself a probability. Popper has thus moved somewhat closer to the logical empiricists while his recent interpreters have moved farther away. Imre Lakatos, for example, refers to his formula for computing degrees of corroboration as "only a curious slip which is out of tune with his philosophy." Imre Lakatos, "History of Science and its Rational Reconstructions," *Boston Studies in the Philosophy Science*, VIII (Reidel, 1971), p. 128.

18 *LSD*, p. 31.

19 *Ibid.*, p. 43.

20 See particularly the work of Imre Lakatos: "Changes in the Problem of Inductive Logic," *The Problem of Inductive Logic*, ed. I. Lakatos (North Holland, 1968); "Falsification and the Methodology of Research Programmes"; "Popper on Demarcation and Induction," *The Philosophy of Sir Karl Popper*, ed. P. A. Schilpp (Open Court, 1974).

21 *LSD*, p. 50. It should be noted that the phrase "or strict disproof" did not appear in the original text but was added to the English edition because, Popper says, he has so often been misinterpreted as holding a doctrine of conclusive falsifiability. *Ibid.*, note *1.

22 See Chapter VII.

23 *LSD*, pp. 43-44.

24 *Ibid.*, p. 104.

25 *Ibid.*, p. 108.

26 *Ibid.*, p. 109.

27 *Ibid.*, p. 86.

28 *Ibid.*

29 *Ibid.*, pp. 86-7. Lakatos consistently misreads this passage as saying that a basic statement which refutes a theory must be supported by a well corroborated falsifying hypothesis ("Falsification and the Methodology of Research Programmes," p. 108 and again on p. 127). This is the reverse of Popper's position.

30 *LSD*, p. 100.

31 "Falsification and the Methodology of Research Programmes."

32 "History of Science and its Rational Reconstructions." According to Lakatos the intellectually honest historian must report evidence which is contrary to his rational reconstruction of history in the footnotes, but he need not let this evidence affect his reconstruction. *Ibid.*, p. 107.

33 *LSD*, p. 50 n. *1.

34 See, for example, A. J. Ayer, *Logical Positivism* (Free Press, 1959), pp. 13-14.

35 "The 'Orthodox' View of Theories," p. 6.

36 See, for example, Norwood Russell Hanson, *Perception and Discovery*, ed. Willard C. Humphreys (Freeman, Cooper, 1969), "Editor's Epilogue," p. 427.

37 Stephen Toulmin, *Human Understanding* (Princeton U. Press, 1972), p. 101.

Chapter
Six

1 Feigl, "The 'Orthodox' View of Theories," p. 5. Feigl attributes this metaphor to Schick, Carnap, Hempel and Margenau.

2 *Perception and Discovery*, p. 61.

3 Pierre Duhem, *The Aim and Structure of Physical Theory*, trans. Philip P. Wiener (Atheneum, 1962), p. 145.

4 Norwood Russell Hanson, *Patterns of Discovery* (Cambridge U. Press, 1958), p. 4.

5 I do not concede the existence of theory-free sense-data, but there is no need to raise that difficult and complex issue here. It is sufficient for our purposes to show that even if there were such data, they could not serve as objects of scientific knowledge nor play any role in resolving scientific disputes.

6 Cf. Hanson, *Patterns of Discovery*, pp. 5-8.

7 Cf. Kordig, *The Justification of Scientific Change*, pp. 16-19 for a recent version of this argument.

8 It is worth noting that it is somewhat self-defeating to attempt to defend the traditional empiricist analysis of scientific perception by appealing to more recent astronomical *theory* to decide what it was that Kepler and Brahe really saw.

9 Thomas S. Kuhn, *The Structure of Scientific Revolutions*, 2d ed. (U. of Chicago Press, 1970), p. 114. Kuhn's purpose in this passage is to show that the gestalt shift and other examples from psychology are merely suggestive and are fundamentally different from cases that arise in science since in the psychological experiment we have an

"external standard" (*Ibid.*) which we can appeal to to show that the same object is being observed throughout, while we have no such external standard in the scientific case. I will argue that we have no such external standard in the case of the geltalt shift either.

10 The following discussion relies heavily on Hanson's analysis of *seeing that*. Cf. *Patterns of Discovery*, ch. 1, esp. pp. 19-30, and *Perception and Discovery*, ch. 6, 7, 8.

11 Thus *seeing as* is a special case of *seeing that*; to see an object as a galvanometer is to see that it is a galvanometer.

12 *Patterns of Discovery*, p. 20.

13 *Ibid.*, p. 59.

14 Cf. my "Perception and Meaning," *American Philosophical Quarterly* Monograph No. 6, 1972, pp. 1-9.

15 *Perception and Discovery*, p. 246.

16 See chapter VIII for a more detailed discussion of the role of accepted theories in the genesis of problems.

17 Above, p. 47.

18 *Logic of Scientific Discovery*, pp. 93-94.

19 Cf. above, pp. 92-93.

20 Above, p. 85.

21 Cf. Richard Rorty, "The World Well Lost," *Journal of Philosophy*, 69, 1972, p. 650.

22 An adequate defense of this approach would require another complete book. I have made a first attempt at developing a theory of perception along these lines in *A Causal Theory of Perception*, unpublished Ph.D. dissertation, Northwestern University, 1970.

23 *Structure of Scientific Revolutions*, p. 118.

24 We will return to the problem of relativism in chapter X.

Chapter Seven

1 E. N. da C. Andrade, *Sir Issac Newton* (Doubleday Anchor, 1954), p. 39.

2 *Rise of Scientific Philosophy*, pp. 101-102.

3 Andrade, *Sir Issac Newton*, pp. 40-41; A. R. Hall, *The Scientific Revolution*, 2d ed. (Beacon Press, 1966), pp. 267-268.

4 Stephen F. Mason, *A History of the Sciences* (Collier Books, 1962), pp. 199-200.

5 Sir Issac Newton, *Principia*, trans. Andrew Motte revised by Florian Cajori (U. of California Press, 1971), p. 147. Much later in *Principia* Newton maintained that he had resolved the problem and discovered that the discrepancy was due to "a cause which I cannot here stop to explain," *Ibid.*, p. 435. Cajori points out that Newton never did provide an explanation even though Pemberton, the editor of the third edition, asked him for "a brief hint" of the solution. *Ibid.*, p. 649.

6 *Ibid.*, p. 650. Cf. also Kuhn, *Structure of Scientific Revolutions*, p. 81.

7 Thomas S. Kuhn, "The Caloric Theory of Adiabatic Compression," *Isis*, XLIX, 1958, pp. 136-137. Cf. also Richard S. Westfall, "Newton

and the Fudge Factor," *Science,* 179, 1973, pp. 752-754. Westfall's article is a fascinating discussion of a number of discrepancies between theory and observation in the first edition of *Principa* and of the way in which Newton manipulated the figures to improve the apparent agreement in the later editions. It should be noted that the gravitational force on the moon is one of the figures that, according to Westfall, Newton doctored.

8 For a detailed account see Morton Grosser, *The Discovery of Neptune* (Harvard U. Press, 1962).

9 Stephen Toulmin and June Goodfield, *The Fabric of the Heavens* (Harper Torchbook, 1961), pp. 254-255.

10 *Structure of Scientific Revolutions,* ch. II-V.

11 Margaret Masterman, "The Nature of a Paradigm," *Criticism and the Growth of Knowledge,* pp. 61-65. Cf. also Dudley Shapere, "The Structure of Scientific Revolutions," *Philosophical Review,* 73, 1964, p. 388; Richard L. Purtill, "Kuhn on Scientific Revolutions," *Philosophy of Science,* 34, 1967, p. 53; Israel Scheffler, *Science and Subjectivity,* p. 88.

12 In his more recent writings Kuhn has drawn a distinction between paradigms as concrete achievements around which a style of scientific research crystallizes and paradigms as the entire set of beliefs, techniques, etc. that a scientific community shares. *Structure of Scientific Revolutions,* p. 175. In the first of the above senses *Principia Mathematica* is the paradigm of logical empiricism. The second sense, which Kuhn now prefers to call the "disciplinary matrix" (*Ibid.*, p. 182), needs to be further divided into an epistemic sense and a sociological sense. The latter refers to the shared commitments that create a scientific community, the former to the role that theories play in guiding research, whether the research of an individual or of a community. It is the epistemic aspect of paradigms that concerns us here. As for the legitimacy of taking research in accordance with a theory as the defining characteristic of normal science, Kuhn, in introducing the term "disciplinary matrix," writes: "Scientists themselves would say they share a theory or a set of theories, and I shall be glad if the term can ultimately be recaptured for this use. As currently used in philosophy of science, however, 'theory' connotes a structure far more limited in nature and scope than the one required here." *Structure of Scientific Revolutions,* p. 182. The "current usage" Kuhn refers to is that of the logical empiricists. It should become abundantly clear that in using the term "theories" here I do not restrict myself to their sense of uninterpreted calculi and correspondence rules.

13 Marshall Clagett, *Greek Science in Antiquity* (Collier Books, 1963), pp. 114-116.

14 E. A. Burtt, *The Metaphysical Foundations of Modern Science* (Doubleday Anchor, 1954), pp. 37-38.

15 Galileo, *Dialogue Concerning the Two Chief World Systems,* trans. Stillman Drake (U. of California Press, 1967), pp. 372-373.

16 Thomas S. Kuhn, *The Copernican Revolution* (Vintage Books, 1957), Ch. 2.

17 Wallace Arthur and Saul K. Fenster, *Mechanics* (Holt, Rinehart, Winston, 1969), p. 386.

18 *Ibid.*, pp. 387-388.

19 In praise of those who upheld the motion of the earth Galileo wrote: "Nor can I ever sufficiently admire the outstanding acumen of those who have taken hold of this opinion and accepted it as true; they have through sheer force of intellect done such violence to their own senses as to prefer what reason told them over that which sensible experience plainly showed them to the contrary." *Dialogue on the Two World Systems*, p. 328. And a bit later, in response to the suggestion that it would have been a great pleasure for Copernicus to see his system confirmed by Galileo's observations with the telescope, "Yes, but how much less would his sublime intellect be celebrated among the learned! For as I said before, we may see that with reason as his guide he resolutely continued to affirm what sensible experience seemed to contradict." *Ibid.*, p. 339. Throughout the *Dialogue* Galileo recognizes that it is Simplicio, the Aristotelian, who has the overwhelming body of experience in his favor. For an extended analysis of this point see Paul K. Feyerabend, "Against Method," *Minnesota Studies in the Philosophy of Science*, IV, pp. 17-130, and "Problems of Empiricism: Part II," *The Nature and Function of Scientific Theories*, ed. Robert G. Colodny (U. of Pittsburgh Press, 1970), pp. 275-353. A detailed critique of Feyerabend's analysis is developed in Peter K. Machamer, "Feyerabend and Galileo: The Interaction of Theories and the Reinterpretation of Experience," *Studies in the History and Philosophy of Science*, 4, 1973, pp. 1-46.

20 Francis Bacon, *The New Organon* (Bobbs-Merrill, 1960), pp. 47-66.

21 *Ibid.*, pp. 131-132. It is an indication of how difficult even Bacon found it to stick to pure observation that his list includes, "Animals, especially and at all times internally; though in insects the heat is not perceptible to touch by reason of the smallness of their size." *Ibid.*

22 Cf. Karl Popper, *Conjectures and Refutations*, pp. 46-47 and *Objective Knowledge*, p. 259.

23 Immanuel Kant, *Critique of Pure Reason*, trans. Norman Kemp Smith (Macmillan, 1963).

24 Kant distinguishes between "inner sense" and "outer sense." We observe events in our own streams of consciousness via inner sense and the objects of inner sense are located only in time, not in space. This distinction need not be pursued for our present purposes.

25 There is a third faculty of the mind, Reason, which provides regulative ideas that guide scientific research even though we have no ground for believing that experienced reality conforms to these ideas. This is not the sort of presupposition which concerns us here. We are only concerned with those presuppositions which are part of science and which are taken as true by scientists.

26 For Kuhn paradigms play the same dual role of structuring both research and the world that the scientist experiences. Cf. "I have so

far argued only that paradigms are constitutive of science. Now I wish to display a sense in which they are constitutive of nature as well." *Structure of Scientific Revolutions,* p. 110.

27 R. G. Collingwood, *An Autobiography* (Oxford U. Press, 1939), ch. V, and *An Essay on Metaphysics* (Oxford U. Press, 1940), ch. V-VII. Collingwood's theory of presuppositions is not nearly so well worked out as Kant's.

28 *Essay on Metaphysics,* p. 32.

29 "The absolute presuppositions of any given society, at any given phase of its history, form a structure which is subject to 'strains' of greater or less intensity which are 'taken up' in various ways, but never annihilated. If the strains are too great, the structure collapses and is replaced by another, which will be a modification of the old with the destructive strain removed. . . ." *Ibid.,* p. 48.

30 R. G. Collingwood, *The Idea of Nature* (Oxford U. Press, 1960), pp. 3-10.

31 Cf. my "Paradigmatic Propositions," *American Philosophical Quarterly,* 12, 1975, pp. 85-90.

32 *Aim and Structure of Physical Theory,* ch. VI.

33 Willard Van Orman Quine, "Two Dogmas of Empiricism," *From a Logical Point of View* (Harper Torchbooks, 1961), p. 41.

34 Cf. my "Notes to the Tortoise," *The Personalist,* 53, 1972, pp. 104-109.

35 *Personal Knowledge,* (Harper Torchbooks, 1964), p. 31.

36 Cf. *Structure of Scientific Revolutions,* ch. X.

Chapter
Eight

1 Aristotle, *On the Heavens,* 296b, W. K. C. Guthrie (Harvard U. Press, 1939), pp. 243-247.

2 Aristotle, *Physics,* 267a, trans. R. P. Hardie and R. K. Gaye, *The Basic Works of Aristotle,* ed. Richard McKeon (Random House, 1941), p. 392. Aristotle's explanation was sufficiently subtle to avoid the objection that he does not explain what keeps the air moving: the power to move objects is imparted by the bowstring to the air "which is naturally adapted for imparting and undergoing motion" (*Ibid.*), so that while each bit of air stops moving as soon as the mover is removed, it is still able to move the next bit of air.

3 Strictly speaking the ancients and medievals used the term "planet" for the celestial objects, including the sun and moon, which move in relation to the fixed stars. This will be of some importance later but for the moment I will continue to use "planet" in the modern sense of the word.

4 Cf. Chapter VII, p. 99.

5 I. Bernard Cohen, *The Birth of a New Physics* (Anchor Books, 1960), p. 39.

6 Stephen Toulmin, *Foresight and Understanding* (Harper Torchbooks, 1961), ch. 3, 4.

7 *Science and Subjectivity,* p. 57.

8 Cf. Galileo, *Dialogue on the Two World Systems,* p. 117; E. J.

183 Dijksterhuis, *The Mechanization of the World Picture*, trans. C. Dikshoorn (Oxford U. Press, 1969), p. 34.

9 *On the Heavens*, 308a, p. 329; 311a, p. 353.

10 *Physics*, 208c, p. 270.

11 *Dialogue on the Two World Systems*, pp. 33-34.

12 A more detailed analysis would show a series of changes in the sense of the concept, with each version having more in common with those between which it falls than with any other version. E.g., for Copernicus the notion of uniform circular motion was still built into the concept of a planet. Galileo held that since all natural motions are circular, there is nothing peculiar to planets in circular motion. Next, reluctantly and with great difficulty, Kepler succeeded in conceiving of planetary motion as non-circular. (For a detailed discussion see Hanson, *Patterns of Discovery*, pp. 74-76.) A continuous series of changes could eventually lead to a situation in which a contemporary version of a concept and its earliest versions have no identifiable common aspect, as is suggested by the following recent example: "As Pioneer 10 rounded Jupiter last week, it sent back spectacular photos of the massive red planet and confirmed that is many ways Jupiter hardly seems like a planet at all." (W. D. Metz, "By Jupiter!," *Science*, 182, 1973, p. 1235.) It is obvious that for Aristotle, Ptolemy, Galileo, Newton or Leverrier the very notion that Jupiter might not be a planet would have been an utter absurdity, on a par with suggesting to Ptolemy that the earth might be a planet. Nor could the evidence mentioned—that Jupiter radiates 2.5 times as much heat as it absorbs, has helium as well as hydrogen in its atmosphere, and has a disc shaped magnetic field—have been relevant to the question of whether Jupiter is a planet for those scientists.

13 *Principia*, p. 1.

14 *Ibid.*, p. 2.

15 Arthur and Fenster, *Mechanics*, p. 5.

16 A. P. French, *Newtonian Mechanics* (Norton, 1971), pp. 164-165.

17 Carl G. Hempel, *Philosophy of Natural Science* (Prentice-Hall, 1966), p. 94.

18 Above, ch. III n. 42.

19 Cf. Norwood R. Hanson, "Logical Positivism and the Interpretation of Scientific Theories," *Legacy of Logical Positivism*, p. 75.

20 The notion of evolution of concepts that I suggest here is different from that of Stephen Toulmin. I am dealing with changes in a concept, and it is not clear that Toulmin recognizes that there is an important respect in which concepts themselves undergo change. Even though volume I of his *Human Understanding* is entitled *The Collective Use and Evolution of Concepts*, what Toulmin studies is the evolution of intellectual disciplines. For him they are constellations of concepts, the development of which is to be understood in terms of the appearance of new concepts and the disappearance of old ones in much the same way that biological species evolve through variation and natural selection. But for Toulmin concepts themselves do not evolve, just as, in the

biological parallel, particular variations do not evolve.
21 *Structure of Scientific Revolutions,* pp. 101-102.
22 Paul K. Feyerabend, "Explanation, Reduction and Empiricism," *Minnesota Studies in the Philosophy of Science,* III, pp. 80-81; "Problems of Empiricism," *Beyond the Edge of Certainty,* ed. Robert Colodny (Prentice-Hall, 1965), pp. 168-170.
23 This has become a standard technique of quantum mechanics under the guidance of the correspondence principle.
24 *Structure of Scientific Revolutions,* p. 102.

Chapter
Nine

1 Richard S. Rudner, *Philosophy of Social Science* (Prentice-Hall, 1966), p. 6.
2 *Experience and Prediction,* p. 6.
3 *Ibid.,* p. 382.
4 *Logic of Scientific Discovery,* p. 31.
5 Florian Cajori, "An Historical and Explanatory Appendix," *Principia,* p. 650.
6 *Philosophy of Natural Science,* p. 14.
7 *Ibid.,* p. 16. Strictly speaking, deductive logic does provide mechanical rules for generating new propositions which follow from those already available. For example, we can formulate a rule which tells us to add some given proposition as a disjunct to any other proposition, but significant theorems are not discovered in this way.
8 Cf. Reichenbach, *Experience and Prediction,* pp. 5-6.
9 *Logic of Scientific Discovery,* p. 28.
10 *Ibid.,* p. 32.
11 This point will be developed further in Chapter X.
12 Cf. Max Jammer's description of the development of quantum mechanics, "Each stage depended on those preceding it without necessarily following from them as a logical consequence." *The Conceptual Development of Quantum Mechanics* (McGraw-Hill, 1966), p. vii.
13 Cf. Errol E. Harris, *Hypothesis and Perception* (Humanities Press, 1970). This is an extended attempt to develop a dialectical analysis of science to which I am greatly indebted. See also my paper, "Harris on the Logic of Science," *Dialectica,* 26, 1972, pp. 227-246.
14 Plato, *Republic,* 331c, trans. F. M. Cornford (Oxford U. Press, 1968), p. 7.
15 *Ibid.*
16 *Ibid.,* 332a-b.
17 *Ibid.,* 332c-336a, pp. 9-14.
18 *Ibid.,* 341b-342e, pp. 22-24.
19 Cf. the discussion of Collingwood above, ch. VII, pp.
20 Cajori, "Appendix," *Principia,* p. 650.
21 In this case we come as close as we can to a situation in which the accepted theory dictates the form the solution to a problem must

185 take. Barring an error in the mathematics (which is what Clairaut eventually discovered in the case of the moon's motion), the only way to explain the unexpected perturbation of Uranus in the context of Newtonian mechanics is by the presence of a hitherto unknown force which could only be exercised by another gravitating body.

22 Note that not even the most ardent advocate of a strict distinction between the context of discovery and the context of verification talks about Leverrier's discovery of Vulcan.

23 Quoted in Kuhn, *The Copernican Revolution,* pp. 138-139.

24 Clagett, *Greek Science in Antiquity,* p. 114.

25 Edward Grant, *Physical Science in the Middle Ages* (Wiley, 1971), pp. 64-70.

26 *Dialogue on the Two World Systems,* p. 339.

27 *Patterns of Discovery,* p. 74.

28 *Ibid.,* pp. 74-76.

29 *Ibid..* pp. 78-83.

30 A. Einstein, "On the Electrodynamics of Moving Bodies," trans. W. Perrett and G. B. Jeffrey, *The Principle of Relativity* (Dover Books), p. 37-38. Italics mine.

31 *Ibid.,* pp. 36, 46.

32 Edmund Whittaker, *A History of the Theories of Aether & Electricity* (Harper Torchbooks, 1960), v. 2, p. 32.

33 Quoted in Ronald W. Clark, *Einstein* (Avon, 1972), p. 59.

34 Quoted in *Ibid.,* p. 113.

35 *History of Theories of Aether & Electricity,* ch. II.

36 Herbert Feigl, "Beyond Peaceful Coexistence," *Minnesota Studies,* V, p. 9.

37 "Electrodynamics of Moving Bodies," p. 37.

38 *Ibid.,* p. 63.

39 *Ibid.,* p. 64.

40 Chapter VIII, pp. 123-126.

41 Cf., for example, Salviati's reply to Simplicio's charge that Plato "plunged into geometry too deeply and became too fascinated by it. After all, Salviati, these mathematical subtleties do very well in the abstract, but they do not work when applied to sensible and physical matters." *Dialogue on Two World Systems,* pp. 203ff.

42 Cf. Stillman Drake, "The Effectiveness of Galileo's Work," *Galileo Studies* (U. Michigan Press, 1970), pp. 95-122. There are still Aristotelian writers who maintain that quantitative science cannot provide explanations of natural phenomena and thus must be supplemented by a non-quantitative "purely physical theory of nature." James A. Weisheipl, *The Development of Physical Theory in the Middle Ages* (U. Michigan Press, 1971), pp. 83-88. In his early work Galileo too accepted a distinction between mathematical elucidation and physical explanation. Cf. *On Motion and On Mechanics,* (U. of Wisc. Press, 1960), p. 20.

43 *Ibid.,* pp. 165-166.

44 *Ibid.,* p. 439.

45 *Ibid.*

46 *Ibid.,* p. 417. For a deatiled defense of this interpretation see my

"Galileo, the Elements, and the Tides," *Studies in the History and Philosophy of Science,* 7, 1976, pp. 337-351.

47 In the Galilean problem situation even theological arguments were relevant, since both Galileo and his opponents accepted the authority of the scriptures as an independent standard to which they had to conform. The problem of providing an acceptable interpretation of scripture from a Copernican viewpoint was taken just as seriously by Galileo as the problem of providing an acceptable explanation of the straight fall of a stone from a tower. Hence Galileo's famous excursion into biblical interpretation in his letter to the Grand Duchess Christina, *Discoveries and Opinions of Galileo,* trans. Stillman Drake (Anchor Books, 1957), pp. 175-216.

Chapter
Ten

1 Plato, *Theatetus,* 152c, trans. F. M. Cornford, *Collected Dialogues,* ed. Edith Hamilton and Huntington Cairns (Pantheon, 1961), p. 857.
2 Cf. Ch. III.
3 Cf. Ch. VI.
4 "Logic of Discovery or Psychology of Research," *Criticism and the Growth of Knowledge,* p. 13.
5 Cf. pp. 131-132.
6 *Structure of Scientific Revolutions,* p. 94.
7 *Ibid.,* p. 148.
8 Aristotle, *Nicomachean Ethics,* 1112a, trans. W. D. Ross, *Basic Works of Aristotle,* p. 969. Aristotle adds that we do not deliberate about the material universe, but this is because he believed that we have necessary knowledge of it too, and that we do not deliberate about things over which we have no control; but this does not concern us here.
9 *Ibid.,* 1094b-1095a, p. 936.
10 *Ibid.,* 1137b, p. 1020.
11 "Reflections on my Critics," *Criticism and the Growth of Knowledge,* pp. 237-238.
12 Cf. above, Ch. VIII, esp. pp. 115-121.
13 Consider the following example: "A few years ago there appeared in *Nature* a table of figures proving with great accuracy that the time of gestation, measured in days, of a number of different animals ranging from rabbits to cows is a multiple of the number π. . . . Yet an exact relationship of this kind makes no impression on the modern scientist and no amount of confirmatory evidence would convince him that there is any relation between the period of gestation of animals and multiples of the number π." Michael Polanyi, *The Logic of Liberty* (U. of Chicago Press, 1951), pp. 16-17. Polanyi's statement is too strong since such a relation might turn out to be significant in some future theory, but the point remains: no matter how firm the correlation may be, it is not a part of science as long as scientists ignore it.
14 Cf. above, pp. 64-65.

15 There is no respect in which our analysis of truth can be considered a version of pragmatism. It is being suggested that a claim is accepted as true because it "works," but "works" here means only that it plays a significant role in the body of scientific knowledge. The pragmatist attempts to reduce theory to practice by defining truth in terms of what works in the practical world; our concern is with theory. General relativity, for example, is accepted by the scientific community and thus true because it allows us to resolve such purely theoretical problems as the computation of the orbit of Mercury, even though the theory has no practical consequences in the sense relevant to pragmatism.

16 *Human Understanding,* p. 111.

17 *Ibid.,* p. 110. Cf. also "Does the Distinction Between Normal and Revolutionary Science Hold Water?" *Criticism and the Growth of Knowledge,* p. 40.

18 Due to lack of funding a number of particle accelerators were closed down in spite of recent discoveries which pose a fundamental challenge to existing theories. Cf. *Science,* 186, 1974, pp. 909-911.

19 *Treatise of Human Nature,* pp. 469-70.

20 Cf. above, p. 28.

21 Cf. above, p. 23.

22 See, for example, *Conjectures and Refutations,* pp. 33-37.

23 For a recent discussion see Loren R. Graham, *Science and Philosophy in the Soviet Union* (Alfred A. Knopf, 1972), ch. 6. A more detailed analysis by a biologist who was personally involved is Zhores A. Medvedev, *The Rise and Fall of T. D. Lysenko,* trans. I. Michael Lerner (Doubleday, 1971).

24 As one example of Lysenko's technique consider the government decree of August 3, 1931 which demanded the development of new strains such as a wheat with, "high yield, uniformity, crystallinity, nonlodging, nonshattering, resistance to cold, drought, pests and disease, good baking quality and other traits . . . in three to four years. . . . Vavilov [the leading Soviet geneticist at the time] viewed the accelerated goals for renewal of seed very sceptically, while Lysenko immediately published a solemn pledge to develop new varieties with preplanned characteristics in two and one-half years." Medvedev, *Rise and Fall of Lysenko,* p. 19.

25 Such occurrences are not limited to Communist societies. During World War II one man, Lord Cherwell, became science advisor to Churchill and thus gained almost complete control over British science. Cf. C. P. Snow, *Science and Government,* (Mentor Books, 1962).

26 We cannot assert that this is always so since we make this judgment from the viewpoint of current science and we have little information about those errors which, thus far, have not been corrected.

27 Michael Polanyi, *Knowing and Being,* ed. Marjorie Grene (U. of Chicago Press, 1969), p. 78.

28 Ralph E. Juergens, "Minds in Chaos," *The Velikovsky Affair,* ed. A.

de Grazia, R. Juergens and L. Stecchini (University Books, 1966), p. 19. This is a partisan pro-Velikovsky book, but there are no non-partisan works on the subject.

29 *Ibid.*, p. 17.

30 *Ibid.*, p. 21.

31 Eric Larabee, "The Day the Sun Stood Still," *Harper's,* 200, Jan. 1950, pp. 19-26. Immanuel Velikovsky, "The Heavens Burst," *Collier's,* Feb. 25, 1950, pp. 24, 42-43, 45, and "World on Fire," March 25, 1950, pp. 25, 82-85. Both of these articles were listed as "excerpted and adapted by John Lear." Fulton Oursler, "Why the Sun Stood Still," *Reader's Digest,* March, 1950, pp. 139-148. *Worlds in Collision* was published on April 3, 1950.

32 Cecilia Payne-Gaposchkin, "Nonsense, Dr. Velikovsky!" *The Reporter,* 2, March 14, 1950, pp. 37-40.

33 *Reporter,* 2, April 11, 1950, p. 2.

34 This is Larabee's own description of his article. *Ibid.*

35 "Nonsense, Dr. Velikovsky!" p. 38.

36 "Minds in Chaos," *Velikovsky Affair,* pp. 35-36. This is one abuse which has been corrected. Although few scientists take Velikovsky's theories seriously, during 1974 he was an invited speaker at meetings of the American Association for the Advancement of Science and at the Philosophy of Science Association, among others.

37 *Ibid.*, p. 26.

38 *Ibid.*, p. 30.

39 *Ibid.*, p. 23.

40 John Q. Stewart, "Disciplines in Collision," *Harper's,* 202, June, 1951, p. 57. The article is part of a debate between Stewart and Velikovsky. Later in his article Stewart offers a summary of Velikovsky's views which is not taken from his own reading of *Worlds in Collision,* but from another review by Payne-Gaposchkin which Stewart himself describes as "definitely unsympathetic." *Ibid.*, p. 59.

41 The following incident will help to underline this. Recently I was reading an exchange of letters between Velikovsky and an astronomy professor in which the latter presented a computation to show that one of the former's claims was incorrect. Velikovsky replied that the astronomer had used the wrong value for a constant and gave instead a value which he supported with a reference, something his opponent had not done. My own reaction was to look at Velikovsky's footnote, and when I saw that the book he cited was reasonably current and was published by Oxford University Press I concluded it was probably reliable.

42 Quoted in "Minds in Chaos," *Velikovsky Affair,* p. 64. Another example from the Lysenko affair is appropriate here. In 1964, after Lysenko's dictatorial power had been broken, many of his followers still remained in powerful positions in Soviet biology. Medvedev, for one, categorically rejected any suggestion that they simply be thrown out: "In 1948 the Lysenkoites achieved a rapid rout of scientific institutions and replacement of editorial boards, academic councils, and so forth, by the basic method of decrees

from ministries, government departments, and boards, and by creation of special plenipotentiary commissions—in other words, by a coup. Today these methods are inapplicable; hence the reverse process is proceeding at a much slower pace. . . .
"They have also been able to utilize the principles brought forward in the struggle against them, and above all the principle of freedom of speech. This permits them now and again, in one form or another, to propagandize their erroneous, false dogmas, to criticize their opponents, and to falsify the real situation in biology. "There is no danger in these activities, which are unavoidable in a democratically structured science." Medvedev, *Rise and Fall of Lysenko,* pp. 242-243.

43 "Minds in Chaos," *Velikovsky Affair,* p. 39.

44 Quoted in *Ibid.*, p. 17. Similarly Philip H. Ableson, editor of *Science,* wrote while explaining why he rejected an article by Velikovsky: "Science can exist and is useful because much of the knowledge in it is more than 99.9 percent certain and reproducible. If science were based on suggestions that were true 50 percent of the time, and all were free to make predictions which were only that reliable, chaos would result. I have repeatedly seen men of brilliance with fertile imaginations make all kinds of suggestions. Ideas are easy. They are cheap. It is the proving of a suggestion beyond a reasonable doubt that makes it valuable." Quoted in de Grazia, "The Scientific Reception System," *Velikovsky Affair,* pp. 188-189.

45 Fred Hoyle, *The Nature of the Universe,* rev. ed. (Signet Books, 1960), pp. 118-119.

46 Fred Hoyle, *Galaxies, Nuclei and Quasars,* (Harper and Row, 1965), p. 131. Italics mine.

47 *Ibid.*, p. 129.

48 Cf. Thomas S. Kuhn, "The Essential Tension," *Third University of Utah Research Conference on the Identification of Creative Scientific Talent,* ed. Calvin W. Taylor (U. of Utah Press, 1959), pp. 341-354.

49 Cf. Stephen G. Brush, "Should the History of Science Be Rated X?" *Science,* 183, 1974, pp. 1164-1172.

Bibliography

d'Abro, A. *The Evolution of Scientific Thought.* New York: Dover, 1950.

_____. *The Rise of the New Physics,* 2 vols. New York: Dover, 1951.

Achinstein, Peter. *Concepts of Science.* Baltimore: John Hopkins U. Press, 1968.

_____. "On the Meaning of Scientific Terms." *Journal of Philosophy,* 61 (1964): 475-510.

Achinstein, Peter and Barker, Stephen F. (eds.). *The Legacy of Logical Positivism.* Baltimore: Johns Hopkins U. Press, 1969.

Agassi, Joseph. *Towards an Historiography of Science.* The Hague: Mouton, 1963.

Andrade, E. N. da C. *Sir Issac Newton.* Garden City: Doubleday, 1954.

Aristotle. *The Basic Works of Aristotle,* ed. Richard McKeon. New York: Random House, 1941.

_____. *On the Heavens,* trans. W. K. C. Guthrie. Cambridge: Harvard U. Press, 1939.

Arthur, Wallace and Fenster, Saul K. *Mechanics.* New York: Holt, Rinehart, Winston, 1969.

Ayer, Alfred Jules. *Language, Truth and Logic.* New York: Dover, 1946.

_____ (ed.). *Logical Positivism.* New York: Free Press, 1959.

Bacon, Francis. *The New Organon.* New York: Bobbs-Merrill, 1960.

Barker, S. F. *Induction and Hypothesis.* Ithaca: Cornell U. Press, 1957.

Barrett, William. "On Dewey's Logic." *Philosophical Review,* 50 (1941): 305-315.

Baumrin, Bernard (ed.). *Philosophy of Science: The Delaware Seminar 1961-62.* New York: John Wiley, 1963.

_____ (ed.). *Philosophy of Science: The Delaware Seminar 1962-63.* New York: John Wiley, 1963.

Blackwell, Richard J. *Discovery in the Physical Sciences.* Notre Dame: U. Notre Dame Press, 1969.

Boas, Marie. *The Scientific Renaissance.* New York: Harper and Row, 1962.

Braithwaite, R. B. *Scientific Explanation.* New York: Harper and Row, 1960.

Bridgman, P. W. *The Logic of Modern Physics.* New York: Macmillan, 1927.

_____. *The Nature of Physical Theory.* Princeton: Princeton U. Press, 1936.

Brodbeck, May. "Explanation, Prediction and 'Imperfect Knowledge,'" Feigl and Maxwell, *Minnesota Studies,* III.

Brown, Harold I. "Harris on the Logic of Science." *Dialectica,* 26 (1972): 227-46.

_____. "Notes to the Tortoise." *Personalist,* 53 (1972): 104-109.

_____. "Paradigmatic Propositions." *American Philosophical Quarterly,* 12 (1975): 85-90.

_____. "Perception and Meaning." *Studies in the Philosophy of Mind,* ed. Nicholas Rescher. Oxford: Basil Blackwell, 1972.

_____. "Problem Changes in Science and Philosophy." Metaphilosophy, 6 (1975): 177-192.

Brush, Stephen G. "Should the History of Science Be Rated X?" *Science,* 183 (1974): 1164-1172.

Buck, Roger C. and Cohen, Robert S. (eds.). *Boston Studies in the Philosophy of Science,* VIII. Dordrecht: D. Reidel, 1971.

Bunge, M. (ed.). *The Critical Approach to Science and Philosophy.* New York: Free Press, 1964.

_____ (ed.). *Delaware Seminar in the Foundations of Physics.* New York: Springer Verlag, 1967.

Burtt, E. A. *The Metaphysical Foundations of Modern Science.* Garden City: Doubleday, 1954.

Butterfield, Herbert. *The Origins of Modern Science,* rev. ed. New York: Free Press, 1957.

Campbell, Norman Robert. *Foundations of Physics.* New York: Dover, 1957.

_____. *What is Science?* New York: Dover, 1952.

Carnap, Rudolph. *The Logical Foundations of Probability.* U. of Chicago Press, 1962.

_____. "The Methodological Character of Theoretical Concepts." Feigl and Scriven, *Minnesota Studies,* I.

_____. "On the Application of Inductive Logic." *Philosophy and Phenomenological Research,* 8 (1947): 133-47.

_____. *Philosophical Foundations of Physics.* New York: Basic Books, 1966.

193   _____. "Testability and Meaning." *Philosophy of Science,* 3
(1936): 419-71, 4 (1937): 1-40.

Clagett, Marshall, (ed.). *Critical Problems in the History of Science.*
Madison: U. of Wisconsin Press, 1959.

_____. *Greek Science in Antiquity.* New York: Collier Books, 1963.

_____. *The Science of Mechanics in the Middle Ages.* Madison: U.
of Wisconsin Press, 1959.

Clark, Ronald W. *Einstein.* New York: Avon, 1972.

Clavelin, Maurice. *The Natural Philosophy of Galileo,* trans. A. J.
Pomerans. Cambridge: M.I.T. Press, 1974.

Cohen, I. Bernard. *The Birth of a New Physics.* Garden City:
Doubleday, 1960.

Cohen, Robert S. and Seeger, Raymond J. *Ernst Mach: Physicist
and Philosopher.* Dordrecht: D. Reidel, 1970.

Cohen, Robert S. and Wartofsky, Marx W., (eds.). *Boston Studies in
the Philosophy of Science,* I. Dordrecht: D. Reidel, 1963.

_____. *Boston Studies in the Philosophy of Science,* II. Dordrecht:
D. Reidel, 1965.

_____. *Boston Studies in the Philosophy of Science,* III. Dordrecht:
D. Reidel, 1968.

_____. *Boston Studies in the Philosophy of Science,* IV. Dordrecht:
D. Reidel, 1969.

_____. *Boston Studies in the Philosophy of Science,* V. Dordrecht:
D. Reidel, 1969.

Collingwood, R. G. *An Autobiography.* Oxford: Oxford U. Press,
1939.

_____. *An Essay on Metaphysics.* Oxford: Oxford U. Press, 1940.

_____. *The Idea of Nature.* New York: Oxford U. Press, 1960.

Colodny, Robert G. (ed.). *Beyond the Edge of Certainty.*
Englewood Cliffs: Prentice-Hall, 1965.

_____. *Frontiers of Science and Philosophy.* Pittsburgh: U. of
Pittsburgh Press, 1962.

_____. *Mind and Cosmos.* Pittsburgh: U. of Pittsburgh Press,
1966.

_____. *The Nature and Function of Scientific Theories.* Pittsburgh:
U. of Pittsburgh Press, 1970.

_____. *Paradigms and Paradoxes.* Pittsburgh: U. of Pittsburgh
Press, 1972.

Conant, James Bryant, (ed.). *Harvard Case Studies in
Experimental Science,* 2 vols. Cambridge: Harvard University
Press, 1966.

Craig, William. "On Axiomatizability Within a System." *Journal of
Symbolic Logic,* 18 (1953): 30-32.

_____. "Replacement of Auxiliary Expressions." *Philosophical
Review,* 65 (1956): 38-55.

Crombie, Alistair, (ed.). *Scientific Change.* New York: Basic Books,
1963.

Dijksterhuis, E. J. *The Mechanization of the World Picture,* trans. C.
Dikshoorn. New York: Oxford U. Press, 1969.

Drake, Stillman. *Galileo Studies.* Ann Arbor: U. of Michigan Press,
1970.

Dreyer, J. L. E. *A History of Astronomy from Thales to Kepler*, 2d ed. New York: Dover, 1953.

Duhem, Pierre. *The Aim and Structure of Physical Theory*, trans. Philip P. Weiner. New York: Atheneum, 1962.

Einstein, A. *Essays in Science*. New York: Philosophical Library, 1934.

———. "On the Electrodynamics of Moving Bodies." *The Principle of Relativity*, trans. W. Perrett and G. B. Jeffery. New York: Dover.

———. *Relativity*. New York: Crown Publishers, 1961.

Einstein, A. and Infeld, Leopold. *The Evolution of Physics*. New York: Simon and Schuster, 1961.

Farrington, Benjamin. *Greek Science*. Middlesex: Penguin Books, 1961.

Feigl, Herbert. "Beyond Peaceful Coexistence." Stuewer, *Minnesota Studies,* V.

———. "Logical Empiricism." *Readings in Philosophical Analysis,* ed. Herbert Feigl and Wilfrid Sellars. New York: Appleton, Century, Crofts, 1949.

———. "The 'Orthodox' View of Theories." Radner and Winokur, *Minnesota Studies,* IV.

———. "Some Major Issues and Developments in the Philosophy of Science of Logical Empiricism." Feigl and Scriven, *Minnesota Studies,* I.

Feigl, Herbert and Brodbeck, May, (eds.). *Readings in the Philosophy of Science.* New York: Appleton, Century, Crofts, 1953.

Feigl, Herbert and Scriven, Michael. *Minnesota Studies in the Philosophy of Science,* I. Minneapolis: U. of Minnesota Press, 1956.

Feigl, Herbert, Scriven, Michael, and Maxwell, Grover. *Minnesota Studies in the Philosophy of Science,* II. Minneapolis: U. of Minnesota Press, 1958.

Feigl, Herbert and Maxwell, Grover. *Minnesota Studies in the Philosophy of Science,* III. Minneapolis: U. of Minnesota Press, 1962.

Feyerabend, Paul K. "Against Method." Radner and Winokur, *Minnesota Studies,* IV.

———. *Against Method.* London: New Left Books, 1975.

———. "Classical Empiricism." *The Methodological Heritage of Newton,* ed. Robert E. Butts. Toronto: U. of Toronto Press, 1970.

———. "Explanation, Reduction, and Empiricism." Feigl and Maxwell, *Minnesota Studies,* III.

———. "On the 'Meaning' of Scientific Terms." *Journal of Philosophy,* 62 (1965): 266-74.

———. "Philosophy of Science: A Subject with a Great Past." Stuewer, *Minnesota Studies,* V.

———. "Problems of Empiricism." Colodny, *Beyond the Edge of Certainty.*

———. "Problems of Empiricism: Part II." Colodny, *The Nature and Function of Scientific Theories.*

195 French, A. P. *Newtonian Mechanics.* New York: W. W. Norton, 1971.

Galileo. "The Assayer." *The Controversy on the Comets of 1618,* trans. Stillman Drake and C. D. O'Malley. Philadelphia: U. of Pennsylvania Press, 1960.

_____. *Dialogue Concerning the Two Chief World Systems,* trans. Stillman Drake. Berkeley: U. of California Press, 1967.

_____. *Dialogues Concerning Two New Sciences,* trans. Henry Crew and Alfonso de Salvio. Evanston: Northwestern U. Press, 1968.

_____. *Discoveries and Opinions of Galileo,* trans. Stillman Drake. Garden City: Doubleday, 1957.

_____. *On Motion and On Mechanics,* trans. I. E. Drabkin and Stillman Drake. Madison: U. of Wisconsin Press, 1960.

_____. *Two New Sciences,* trans. Stillman Drake. Madison: U. Wisconsin Press, 1974.

Geymonat, Ludivico. *Galileo Galilei,* trans. Stillman Drake. New York: McGraw-Hill, 1965.

Gillispie, Charles Coulton, *The Edge of Objectivity.* Princeton: Princeton U. Press, 1960.

Goodman, Nelson. *Fact, Fiction, and Forecast.* New York: Bobbs-Merrill, 1965.

Graham, Loren R. *Science and Philosophy in the Soviet Union.* New York: Alfred A. Knopf, 1972.

Grant, Edward. *Physical Science in the Middle Ages.* New York: John Wiley, 1971.

de Grazia, A., Juergens, R., and Stecchini, L. *The Velikovsky Affair.* New York: University Books, 1966.

Grosser, Morton. *The Discovery of Neptune.* Cambridge: Harvard U. Press, 1962.

Hall, A. R. *From Galileo to Newton.* New York: Harper and Row, 1963.

_____. *The Scientific Revolution,* 2d ed. Boston: Beacon Press, 1966.

Hallam, A. *A Revolution in the Earth Sciences.* Oxford: Oxford U. Press, 1973.

Hanson, Norwood Russell. *The Concept of the Positron.* Cambridge: Cambridge U. Press, 1963.

_____. *Observation and Explanation.* New York: Harper and Row, 1971.

_____. *Patterns of Discovery.* Cambridge: Cambridge U. Press, 1958.

_____. *Perception and Discovery,* (ed.) Willard C. Humphreys. San Francisco: Freeman, Cooper, 1969.

Harré, Rom. *An Introduction to the Logic of the Sciences.* New York: Macmillan, 1967.

_____. *The Principles of Scientific Thought.* Chicago: U. of Chicago Press, 1970.

Harris, Errol E. "Epicyclic Popperism." *British Journal for the Philosophy of Science,* 23 (1972): 55-67.

_____. *Hypothesis and Perception.* London: George Allen & Unwin, 1970.

Hempel, Carl G. *Aspects of Scientific Explanation*. New York: Free Press, 1965.
_____. "Deductive-Nomological vs. Statistical Explanation." Feigl and Maxwell, *Minnesota Studies*, III.
_____. *Fundamentals of Concept Formation in Empirical Science*. Chicago: U. of Chicago Press, 1952.
_____. "On the 'Standard Conception' of Theories." Radner and Winokur, *Minnesota Studies*, IV.
_____. *Philosophy of Natural Science*. Englewood Cliffs: Prentice-Hall, 1966.
_____. "A Purely Syntactical Definition of Confirmation." *Journal of Symbolic Logic*, 8 (1943): 122-43.
_____. "Studies in the Logic of Confirmation." *Mind*, 54 (1945): 1-26, 97-121.
Hempel, Carl G. and Oppenheim, Paul. "Studies in the Logic of Explanation." *Philosophy of Science*, 15 (1948): 135-175.
Hesse, Mary B. *Forces and Fields*. Totowa: Littlefield, Adams, 1965.
_____. *Models and Analogies in Science*. Notre Dame: U. of Notre Dame Press, 1970.
_____. *The Structure of Scientific Inference*. Berkeley: U. of California Press, 1974.
Holton, Gerald. *Thematic Origins of Scientific Thought*. Cambridge: Harvard U. Press, 1973.
Hooker, C. A. "Empiricism, Perception and Conceptual Change." *Canadian Journal of Philosophy*, 3 (1973): 59-75.
_____. "On Global Theories." *Philosophy of Science*, 42 (1975): 152-179.
_____. "Philosophy and Metaphilosophy of Science: Empiricism, Popperianism and Realism." *Synthese*, 32, (1975): 177-231.
Hoyle, Fred. *Galaxies, Nuclei and Quasars*. New York: Harper and Row, 1965.
_____. *The Nature of the Universe*, rev. ed. New York: Signet Books, 1960.
Hume, David. *An Enquiry Concerning Human Understanding*, 2d ed., (ed.) L. A. Selby-Bigge. Oxford: Oxford U. Press, 1967.
_____. *A Treatise of Human Nature*, (ed.) L. A. Selby-Bigge. Oxford: Oxford U. Press, 1966.
Humphreys, Willard C. *Anomalies and Scientific Theories*. San Francisco: Freeman, Cooper & Company, 1968.
Jammer, Max. *The Conceptual Development of Quantum Mechanics*. New York: McGraw-Hill, 1966.
_____. *Concepts of Force*. Cambridge: Harvard U. Press, 1957.
_____. *Concepts of Mass*. Cambridge: Harvard U. Press, 1961.
_____. *Concepts of Space*, 2d ed. Cambridge: Harvard U. Press, 1970.
Joergensen, Joergen. *The Development of Logical Empiricism*. Chicago: U. of Chicago Press, 1951.
Kant, Immanuel. *Critique of Pure Reason*, trans. Norman Kemp Smith. New York: Macmillan, 1963.

————. *Metaphysical Foundations of Natural Science,* trans. James Ellington. New York: Bobbs-Merrill, 1970.

Kisiel, Theodore and Johnson, Galen. "New Philosophies of Science in the USA." *Zeitschrift fur allgemeine Wissenschaftstheorie,* 5 (1974): 138-191.

Kordig, Carl R. *The Justification of Scientific Change.* New York: Humanities Press, 1971.

Koyre, Alexandre. *Etudes Galileennes.* Paris: Hermann, 1966.

————. *From the Closed World to the Infinite Universe.* Baltimore: Johns Hopkins U. Press, 1957.

————. *Metaphysics and Measurement.* Cambridge: Harvard U. Press, 1968.

————. *Newtonian Studies.* Chicago: U. of Chicago Press, 1965.

Kuhn, Thomas S. "The Caloric Theory of Adiabatic Compression." *Isis,* 49 (1958): 132-40.

————. *The Copernican Revolution.* New York: Vintage Books, 1957.

————. "The Essential Tension." *Third University of Utah Research Conference on the Identification of Creative Scientific Talent,* (ed.) Calvin W. Taylor. Salt Lake City: U. of Utah Press, 1959.

————. "Notes on Lakatos." Buck and Cohen, *Boston Studies,* VIII.

————. *The Structure of Scientific Revolutions,* 2d ed. Chicago: U. of Chicago Press, 1970.

Lakatos, Imre. "Changes in the Problem of Inductive Logic." *The Problem of Inductive Logic,* (ed.) I. Lakatos. Amsterdam: North-Holland Publishing, 1968.

————. "History of Science and its Rational Reconstructions." Buck and Cohen, *Boston Studies,* VIII.

————. "Proofs and Refutations." *British Journal for the Philosophy of Science,* 14 (1963): 1-25, 120-39, 221-45, 296-342.

Lakatos, I. and Musgrave, A., (eds.). *Criticism and the Growth of Knowledge.* Cambridge: Cambridge U. Press, 1970.

Leonard, Henry S. "Review of Rudolph Carnap, 'Testability and Meaning.' " *Journal of Symbolic Logic,* 2 (1937): 49-50.

Mach, Ernst. *The Science of Mechanics,* trans. Thomas J. McKormack. LaSalle: Open Court, 1960.

Machamer, Peter K. "Feyerabend and Galileo: The Interaction of Theories and the Reinterpretation of Experience." *Studies in the History and Philosophy of Science,* 4 (1973): 1-46.

Mason, Stephen F. *A History of the Sciences.* New York: Collier Books, 1962.

Medvedev, Zhores A. *The Rise and Fall of T. D. Lysenko,* trans. I. Michael Lerner. Garden City: Doubleday, 1971.

Mises, Richard von. *Positivism.* Cambridge: Harvard U. Press, 1951.

Naess, Arne. *The Pluralist and Possibilist Aspects of the Scientific Enterprise.* Oslo: Universitetsforlaget, 1972.

Nagel, Ernest. "The Meaning of Reduction in the Natural Sciences." *Philosophy of Science,* eds. Arthur Danto and Sidney Morgenbesser. New York: Meridian Books, 1960.

_____. *The Structure of Science*. New York: Harcourt, Brace, and World, 1961.

Newton, Issac. *Mathematical Principles of Natural Philosophy,* 2 vols., trans. Andrew Motte, rev. Florian Cajori. Berkeley: U. of California Press, 1971.

_____. *Optics*. New York: Dover, 1952.

Palter, Robert M., ed. *Toward Modern Science*. New York: E. P. Dutton, 1969.

Partington, James R. *A Short History of Chemistry*. New York: Harper and Row, 1960.

Payne-Gaposchkin, Cecilia. "Nonsense, Dr. Velikovsky!" *Reporter,* 2 (March 14, 1950): 37-40.

Pearce, Glenn and Maynard, Patrick, eds. *Conceptual Change*. Dordrecht: D. Reidel Publishing Company, 1973.

Plato. *Collected Dialogues,* ed. Edith Hamilton and Huntington Cairns. New York: Pantheon, 1961.

_____. *Republic,* trans. F. M. Cornford. New York: Oxford U. Press, 1945.

Polanyi, Michael. *Knowing and Being,* ed. Marjorie Grene. Chicago: U. of Chicago Press, 1969.

_____. *The Logic of Liberty.* Chicago: U. of Chicago Press, 1951.

_____. *Personal Knowledge.* New York: Harper and Row, 1964.

_____. *Science, Faith and Society.* Chicago: U. of Chicago Press, 1946.

_____. *The Tacit Dimension.* Garden City: Doubleday, 1967.

Popper, Karl. *Conjectures and Refutations.* New York: Harper and Row, 1968.

_____. *The Logic of Scientific Discovery. New York: Harper and Row, 1959.

_____. *Objective Knowledge.* New York: Oxford U. Press, 1972.

Purtill, Richard L. "Kuhn on Scientific Revolutions." *Philosophy of Science,* 34 (1967): 53-8.

Putnam, Hilary. "The Analytic and the Synthetic." Feigl and Maxwell, *Minnesota Studies,* III.

_____. "What Theories are Not." *Logic, Methodology and Philosophy of Science,* ed. Ernest Nagel, Patrick Suppes, Alfred Tarski. Palo Alto: Stanford U. Press, 1962.

Quine, Willard Van Orman. *From a Logical Point of View.* New York: Harper and Row, 1961.

Radner, Michael and Winokur, Stephen, (eds.). *Minnesota Studies in the Philosophy of Science,* IV. Minneapolis: U. of Minnesota Press, 1970.

Ramsey, F. P. *The Foundations of Mathematics.* Patterson: Littlefield, Adams, 1960.

Ravetz, Jerome R. *Scientific Knowledge and its Social Problems.* New York: Oxford U. Press, 1971.

Reichenbach, Hans. *Experience and Prediction.* Chicago: U. of Chicago Press, 1938.

_____. *The Rise of Scientific Philosophy.* Berkeley: U. of California Press, 1966.

Rorty, Richard. "The World Well Lost." *Journal of Philosophy*, 67 (1972): 649-65.

Rudner, Richard S. *Philosophy of Social Science*. Englewood Cliffs: Prentice-Hall, 1966.

Russell, Bertrand. *Introduction to Mathematical Philosophy*. London: George Allen and Unwin, 1919.

_____. *Mysticism and Logic*. Garden City: Doubleday, 1957.

_____. *Principles of Mathematics*, 2d ed. New York: W. W. Norton, 1937.

Schaffner, Kenneth F. "Outlines of a Logic of Comparative Theory Evaluation with Special attention to Pre- and Post-Relativistic Electrodynamics." Stuewer, *Minnesota Studies*, V.

Scheffler, Israel. *The Anatomy of Inquiry*. New York: Bobbs-Merrill, 1963.

_____. *Science and Subjectivity*. New York: Bobbs-Merrill, 1967.

Schilpp, Paul Arthur, (ed.). *Albert Einstein: Philosopher-Scientist*. LaSalle: Open Court, 1949.

_____, (ed.). *The Philosophy of Karl Popper*. LaSalle: Open Court, 1974.•

_____, (ed.). *The Philosophy of Rudolph Carnap*. LaSalle: Open Court, 1963.

Scriven, Michael. "Explanation and Prediction in Evolutionary Theory." *Science*, 130 (1959): 477-82.

_____. "Explanations, Predictions, and Laws." Feigl and Maxwell, *Minnesota Studies*, III.

Sellars, Wilfrid. *Science, Perception and Reality*. London: Routledge and Kegan Paul, 1963.

Shapere, Dudley. *Galileo*. Chicago: U. of Chicago Press, 1974.

_____. "Meaning and Scientific Change." Colodny, *Mind and Cosmos*.

_____. "The Paradigm Concept." *Science*, 172 (1971): 706-9.

_____, (ed.). *Philosophical Problems of Natural Science*. New York: Macmillan, 1965.

_____. "The Structure of Scientific Revolutions." *Philosophical Review*, 73 (1964): 383-94.

Shea, William R. "Beyond Logical Empiricims." *Dialogue*, 10 (1971): 223-42.

_____. *Galileo's Intellectual Revolution*. London: Macmillan, 1972.

Snow, C. P. *Science and Government*. New York: Mentor Books, 1962.

Stewart, John Q. "Disciplines in Collision." *Harper's*, 202 (June, 1951): 57-63.

Stuewer, Roger H., (ed.). *Minnesota Studies in the Philosophy of Science*, V. Minneapolis: U. of Minnesota Press, 1970.

Suppe, Frederick, (ed.). *The Structure of Scientific Theories*. Urbana: U. of Illinois Press, 1974.

Swinburne, R. G. "The Paradoxes of Confirmation—A Survey." *American Philosophical Quarterly*, 8 (1971): 318-30.

Toulmin, Stephen. *Foresight and Understanding*. New York: Harper and Row, 1961.

_____. *Human Understanding.* Princeton: Princeton University Press, 1972.

_____. *The Philosophy of Science.* New York: Harper and Row, 1953.

_____. (ed.). *Physical Reality.* New York: Harper and Row, 1970.

Toulmin, Stephen and Goodfield, June. *The Fabric of the Heavens.* New York: Harper and Row, 1961.

Velikovsky, Immanuel. *Worlds in Collision.* New York: Macmillan, 1950.

Wartofsky, Marx W., ed. *Boston Studies in the Philosophy of Science,* I. Dordrecht: D. Reidel, 1963.

_____. *Conceptual Foundations of Scientific Thought.* New York: Macmillan, 1968.

Weinberg, Julius R. *An Examination of Logical Positivism.* New York: Harcourt, Brace and World, 1936.

Weisheipl, James A. *The Development of Physical Theory in the Middle Ages.* Ann Arbor: U. of Michigan Press, 1971.

Westfall, Richard S. *The Construction of Modern Science.* New York: John Wiley, 1971.

_____. "Newton and the Fudge Factor." *Science,* 179 (1973): 751-8.

Whittaker, Edmund. *A History of Theories of Aether and Electricity,* 2 vols. New York: Harper and Row, 1960.

Wilson, J. Tuzo, (ed.). *Continents Adrift.* San Francisco: W. H. Freeman, 1972.

Wittgenstein, Ludwig. *Tractatus Logico-Philosophicus,* trans. D. F. Pears and B. F. McGuiness. London: Routledge and Kegan Paul, 1961.

Whitehead, Alfred North and Russell, Bertrand. *Principia Mathematica,* 3 vols. Cambridge: Cambridge U. Press, 1910-1913.

Yourgrau, Wolfgang, (ed.). *Physics, Logic and History.* New York: Plenum Press, 1970.

Ziman, John. *Public Knowledge.* Cambridge: Cambridge U. Press, 1968.

Index